王昀

60平米极小城市

中国建筑工业出版社

左图：60m²极小城市所在的北京大学教职工宿舍楼的外观及周边环境。其中画线部分是60m²极小城市的所在地

Figure Left: 60m² Minimal City's location, the residence community surroundings at Peking University. The underline part is where 60m² Minimal City located

自序 Preface

2001年到北京大学任教时，学校借给我一套60m²的房子，房子位于圆明园西侧北京大学专为北大教职工建设的一个住宅小区内，是一栋6层、砖混结构、一梯两户的住宅楼。

过去的人们从外面回家，经历的空间序列大多是这样的：“宽马路→小巷→进院→进家门”，在这个过程中，人的心理尺度伴随着城市尺度的递减而递减，反过来从“家→城市”，人的心理尺度伴随着城市尺度的递增而递增。随着城市尺度的扩大，街道一而再、再而三地拓宽，这种人与城市的相互关系，逐渐地已不复存在。现在的人们，从家里出来，经过楼梯的三绕两绕，打开楼门立即便是两座楼房之间的空地了，接着就行走在“广场”般尺度的街道上……这种空间尺度忽强忽弱的强烈反差，使人们心理的尺度变化从以往的逐渐递增或逐渐递减的变化状态转变成当下的“大起大落”。

事实上现在我们的住宅室内的布局设计与现在城市的状况非常相似。比如租借给我的这套住宅，其内部各个房间之间的关系同样没有任何“渐变”与“过渡”。

如何让住宅本身与自身的生活密切关联？这本小册子便是当时针对这个60m²房子内部进行重新规划改造时的部分片段与思考的呈现。

When I came to teach in Peking University in 2001, the university lent me a 60m² apartment. It was located in the residence community on the west side of the Summer Palace, which was especially built for the faculties in Peking University. The apartment building is brick-concrete-structured with 6 floors, and it has two households entering from one elevator.

In the past, people would usually pass a routine of different street views to go back: street/avenue — alley — entrance of courtyard — home. During this process, people's psychological measurement decreases along with the city, and vice versa. However, this relationship between human and city has disappeared because of the consistent enlargement of the scale of cities and streets. Nowadays, passing the stairs, you will find yourself standing between two buildings, and walking on the streets which have the same size as squares. This acute contrast of space has shaped peoples' minds from gradually decrease or increase to thrive or fail dramatically.

In fact, my interior design is somewhat similar to the layout of modern city. For example, the condo I borrowed has no transitions between each rooms.

"How to closely relate living spaces to human life" is the main thing this booklet is going to present, which is also considered as renovating this 60m² apartment.

王　昀

目录

左图：作者在他所设计的60㎡极小城市的“广场”上休闲阅读

Figure Left: The author is reading at the “square” in his 60m² Minimal City design

图：60㎡极小城市的入口
gure Left: Entrance of
e 60m^2 Minimal City

引子

60m^2 极小城市的设计是在 2001 年我刚回国时完成的。当时北京大学租借给我一套使用面积为 60m^2 的单元住宅，以作为归国人员的周转房来使用。住宅位于北京圆明园西墙外骚子营附近的北大教师宿舍区一栋砖混结构的单元楼内，共 6 层，一梯两户，看上去大约是 20 世纪 90 年代末修建的。住宅的户型为 60m^2 的使用面积，被设计成三室一厅的格局。住宅南向有两间卧室，每间约 11.5m^2。北向有一小间，面积大约 7.35m^2，住宅的中间位置布置的是起居室和厨房。起居室中一部分墙面是面向北侧的，因而能够获得一个可直接朝北采光的窗。

走进住宅，发现这个所谓的三室一厅，实际上也仅仅就是划分出了三个房间和一个厅而已，住宅内卫生间的门直接对着起居室开启，是一件不能忍受的事情。从功能的角度上来谈，有卫生间和让卫生间用起来方便和好用实际上完全是两码事。

Preliminary

The design of this 60m^2 minimal city was finished in 2001, not very long after I came back to China. As a temporary residency, a 60m^2 apartment was lent for me. It was located in a faculty-residenter building, which is brick-concrete-structured. The building has 6 floors with two-units-on-one-lift style. It looked like to be built in the late 1990s. The usable area of this apartment is 60m^2 with 3 bedrooms and a living room. Among the three bedrooms, two face south, each has 11.5m^2. The one facing north is relatively small, which has 7.35m^2. In the middle of this apartment are arranged as living room and kitchen. A wall of living room faces north, thus it has a window which could directly receive lighting from the north.

As walking into this apartment, you will recognise the layout is carelessly designed. The door of restroom is right facing the living room, which is hard to tolerate. From the perspective

左图："广场"一角

Figure Left: One corner of the 60m^2 Minimal City

此外，厨房的设计也存在很大问题，关键是不知道餐桌怎么摆。就是说在这套住宅内各个功能房间之间的关系以及布局设计等，实际上都与自己未来生活的展开有非常大的矛盾。因此结论是：改造不可避免。但当你想进行改造时，又会发现这个住宅的结构所带来的改造方面的困难是巨大的。因为这个住宅楼的结构采用的是砖混的墙体承重墙结构，结构墙本身又使得户型内部无法进行调整。

既然是自己即将生活在这个空间里，自然就要设法使其满足自己的居住意愿和生活需求。居住空间不仅仅是一个使用面积的问题，更重要的还有：如何让住宅中的空间在适应自己需要的同时，还可变得有趣，在其内部展开生活时，能够感受到空间处在一个不断扩展的状态……这一切居然成为开始设计的动机。

我在思考和勾画平面图的过程中，慢慢地开始试图将自己的家变成一个可带有记忆性，或能唤起自己记忆的一个场所。

经验与设计

至2001年底，在回国之前的10年间，我游走了不少的聚落和城市。在这个旅行的过程中，很多特别精彩的聚落和城市的景象不自觉地印刻在头脑里，成为挥之不去、时而闪过的影子。

在进行这个家的设计过程中，不经意间眼前会浮现出曾经去过和调查过的城市和聚落中的片段和场景。而这些场景本身也自然地于自觉与不自觉之间流向笔端。

"家"，虽然其物理空间相对很小，但在空间的构

of functioning, "having a restroom" and "having a convenient restroom" are two completely different concepts.

Besides that, the design of the kitchen also had large space of improvement. The main problem is how to set the dining table. Actually, the relationship and the layout of each functioning rooms will have a lot of contractions with my life here in this apartment. Therefore, it is unavoidable to renovate. However, when carrying out the transformation, you will discover the tremendous difficulties caused by built-in structure. Because this building is structured by brick-concrete bearing walls, the inner layout is unchangeable.

Since I was going to live in this space, it has to satisfy my will and living requirements. The living space isn't all about usable area, the more important thing is how to turn your living area into an interesting space with an ever-expanding status, while adapting all kinds of basic needs. And this had become my motivation of design.

In the process of conceiving and drawing the plan, I was gradually trying to transform my home into a place with memories or a place that could remind me of many memories.

Experiences and Design

Until I came back to China in late 2001, I had traveled to large number of communities and cities in 10 years. Some of them have left deep impressions in my mind. I often recall the beauties of their designs.

When designing this home, without any intention, I included some of the elements of the cities and communities I had been to into my design of plan.

Although "home" has relatively small physical space, it

60m²极小城市轴测
The axonometric drawing of the 60m² Minimal Cit

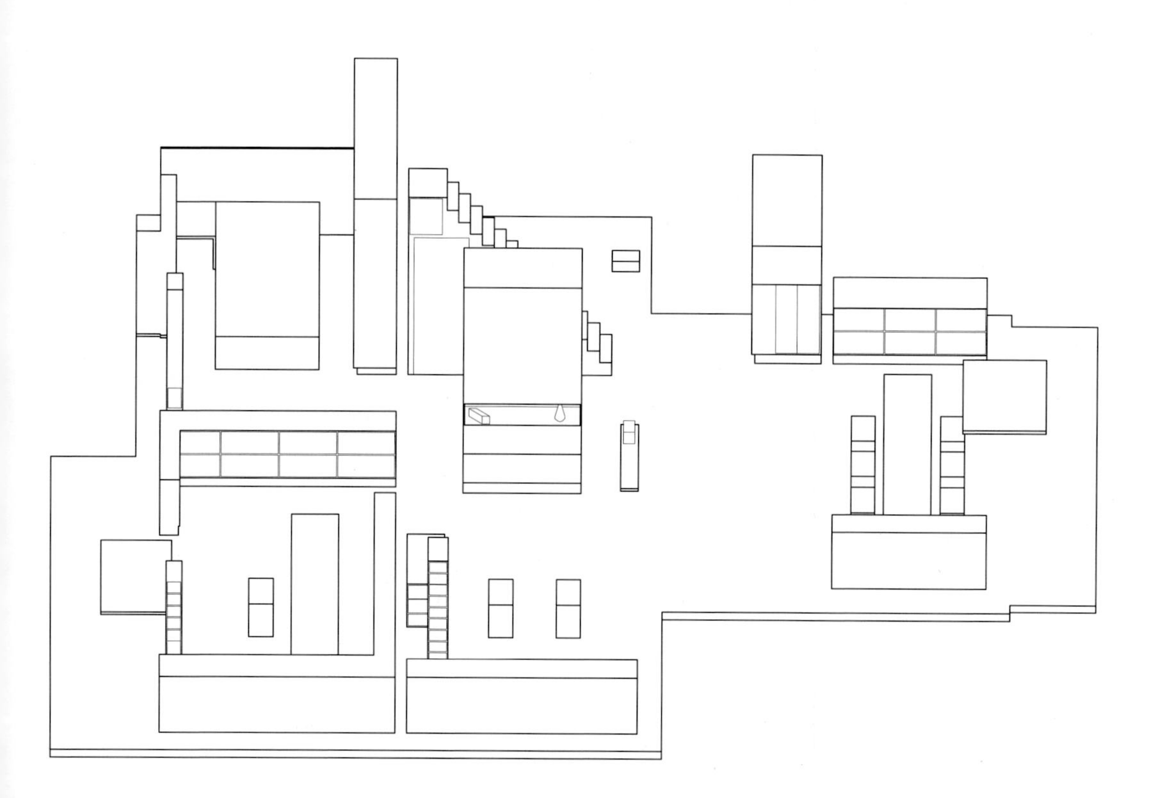

0m²极小城市透视图

The perspective drawing of the 60m² Minimal City

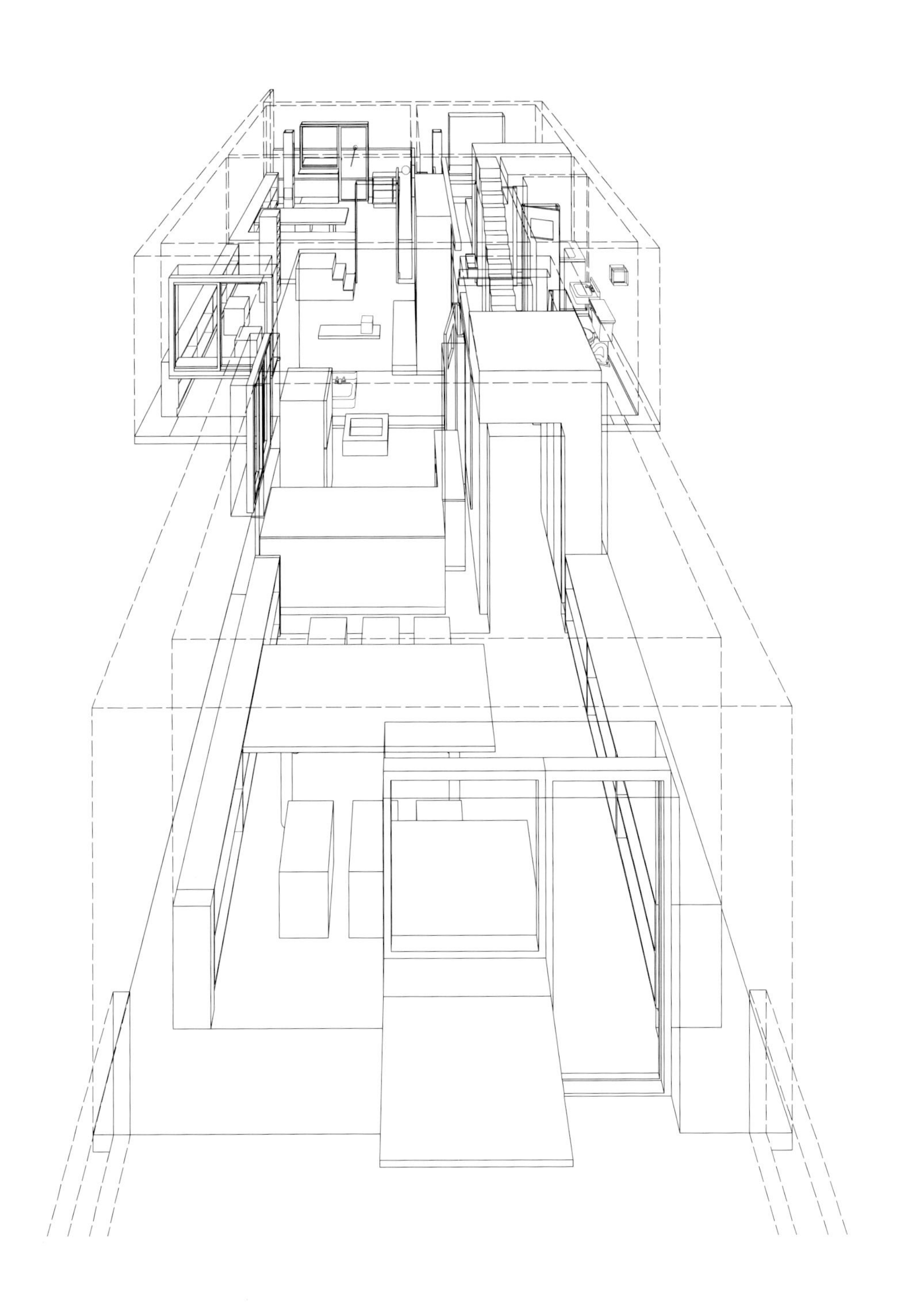

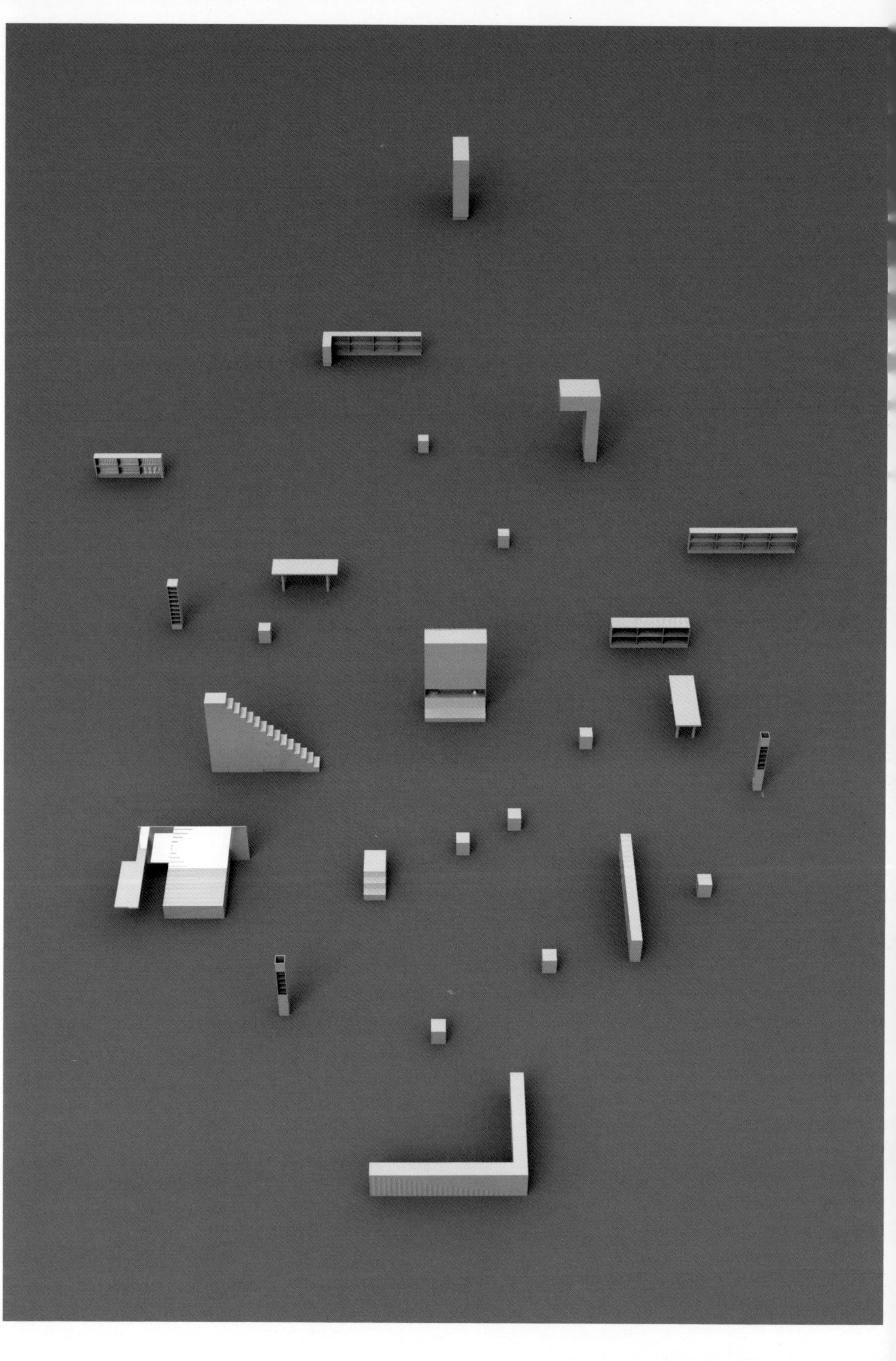

左图：家具在平面中的布局图
右图：极小城市的平面图
Figure Left: Furniture layout plan
Figure Right: The ichnography of the 60m² Minimal City

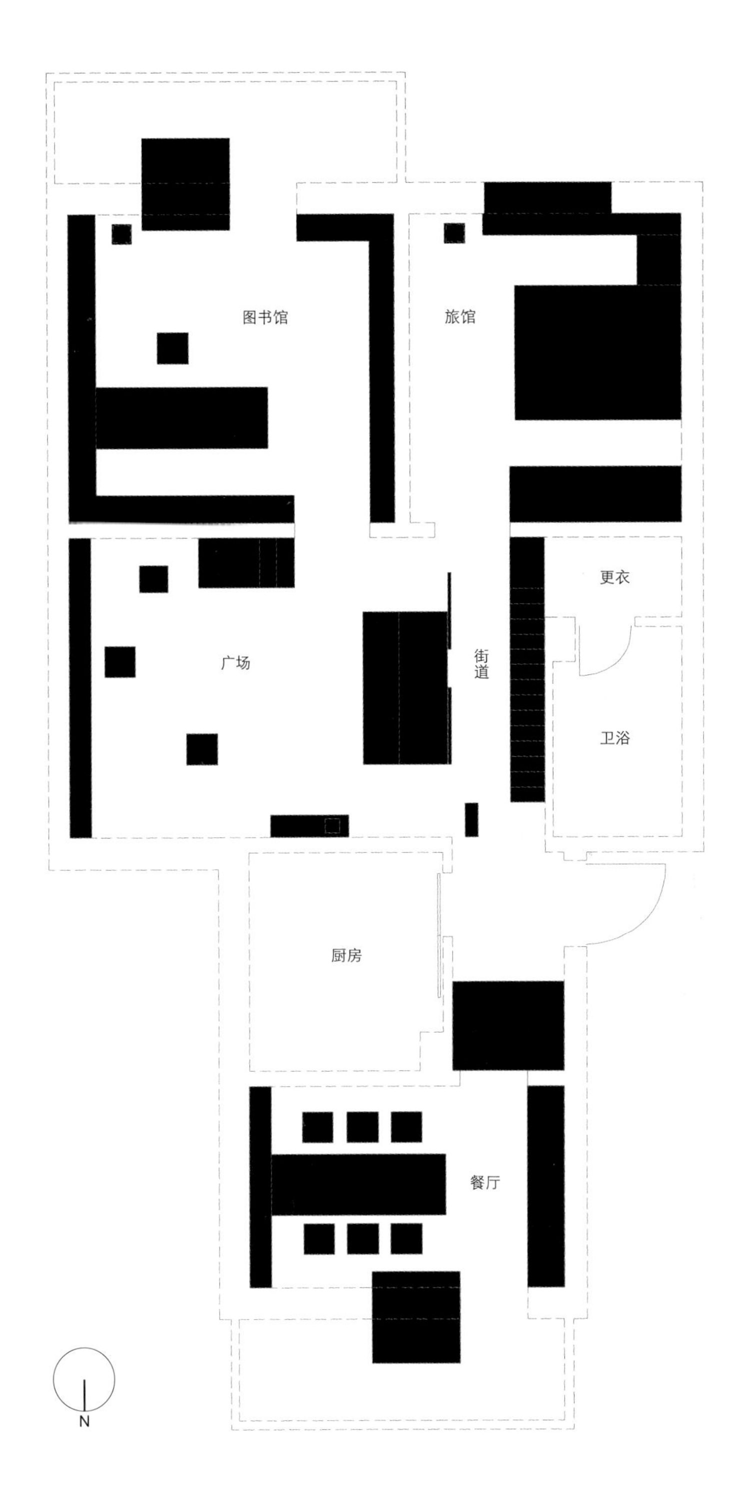

The space transparent relations at square surrounding in the minimal city

“极小城市”中广场周边的透明关系

The space transparent relations of south side of the square

由广场向南侧的空间透明关系

The space transparent relations of the north direction from south side of the square

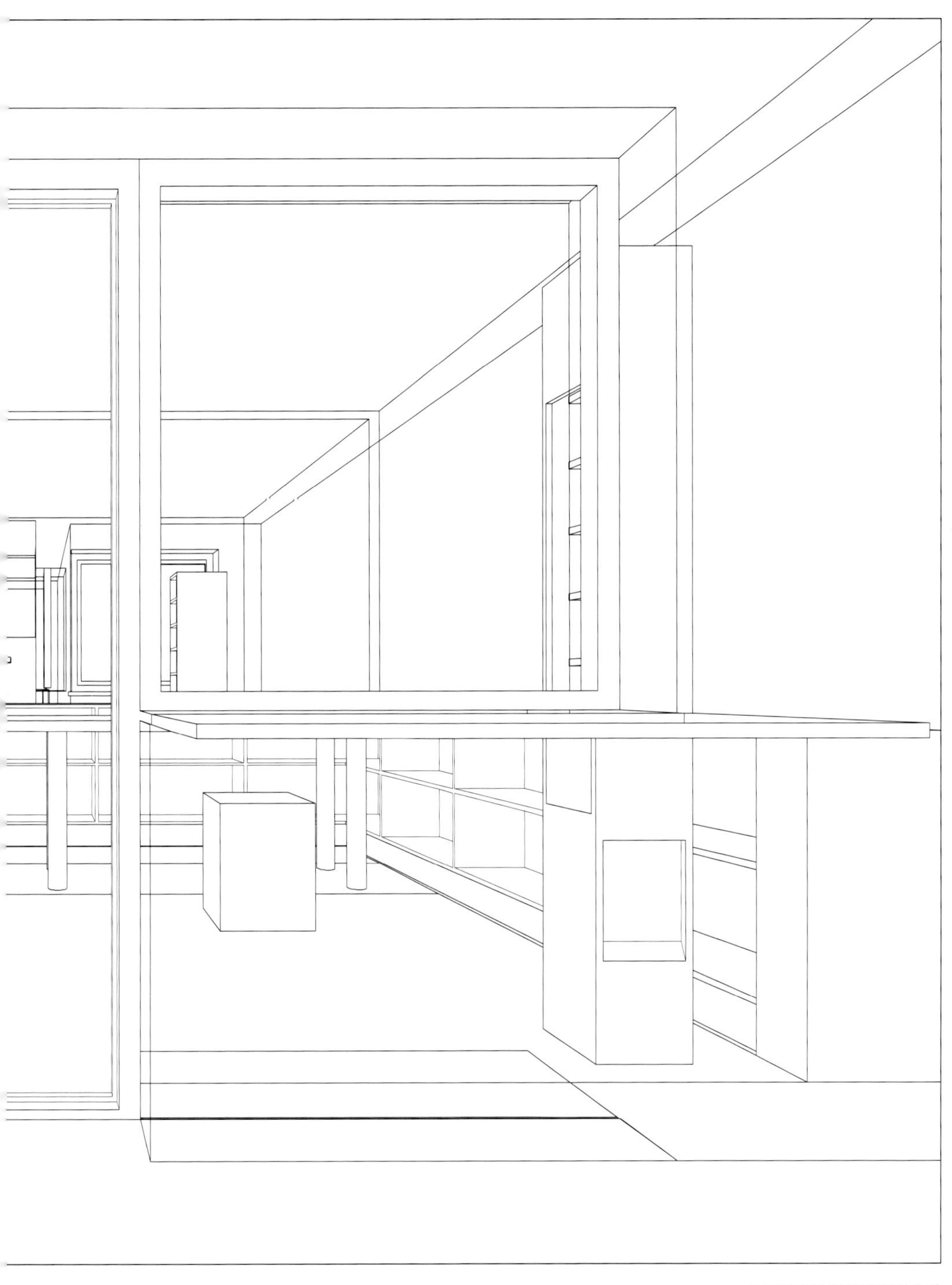

从南侧向北的空间透明关系

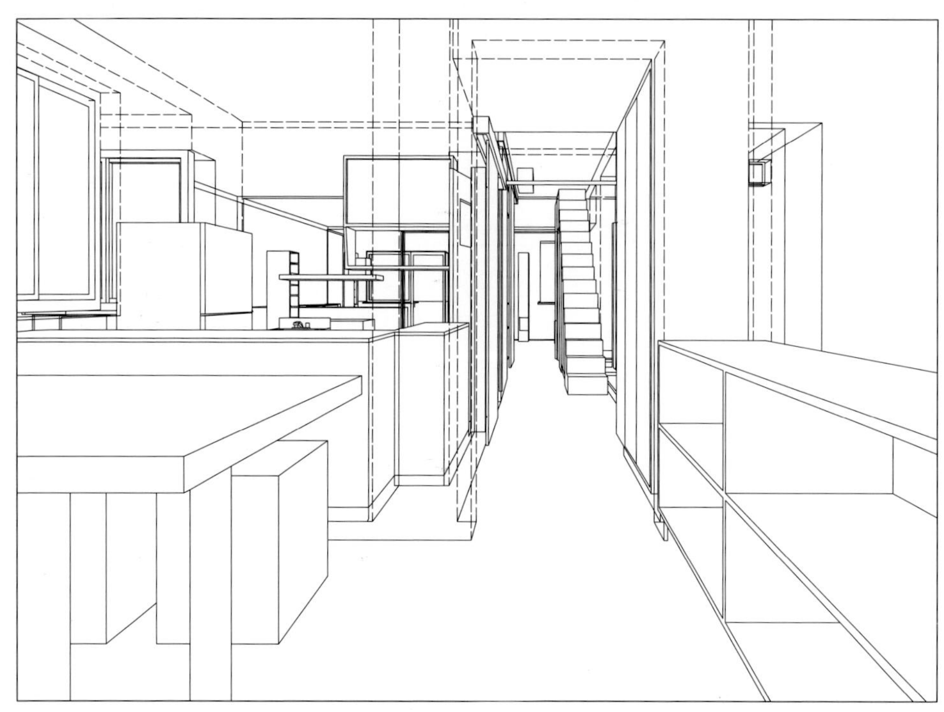

从餐厅向南的空间透明关系
The space transparent relations of the south direction from dining room

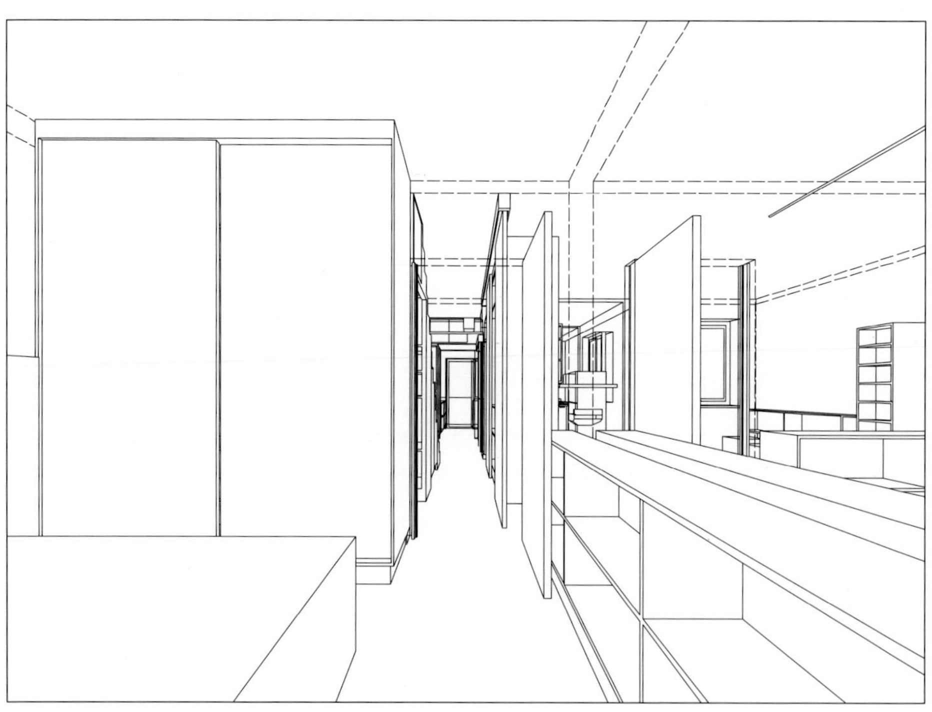

从卧室看向街道的空间透明关系
The space transparent relations from the bedroom view facing to the street

门厅周边的空间透明关系
The space transparent relations of the hallway surrounding

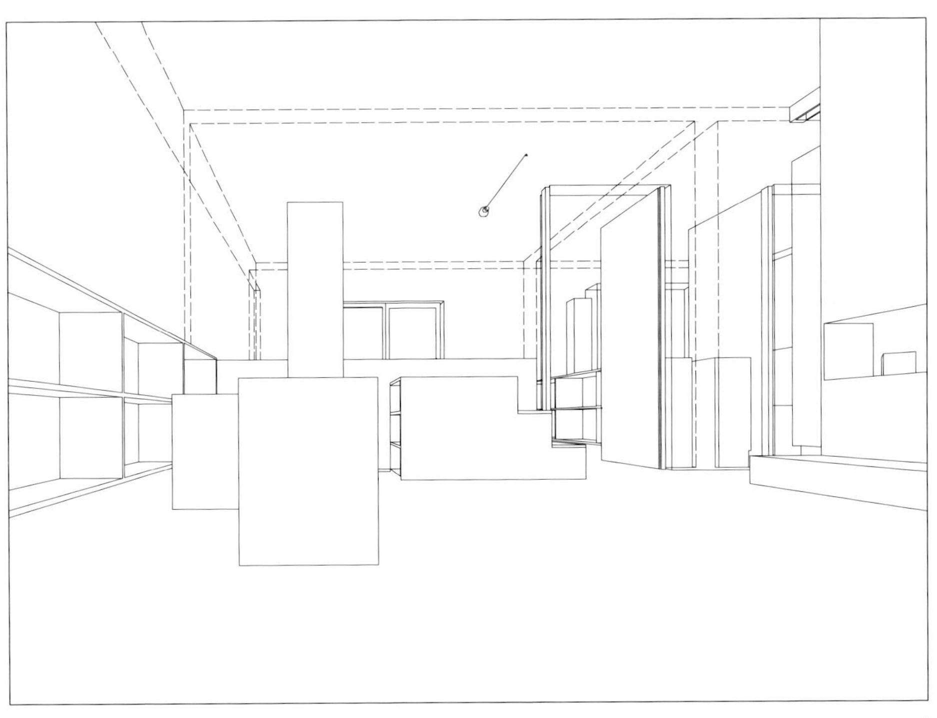

广场与书房的空间透明关系
The space transparent relations of the square and the study

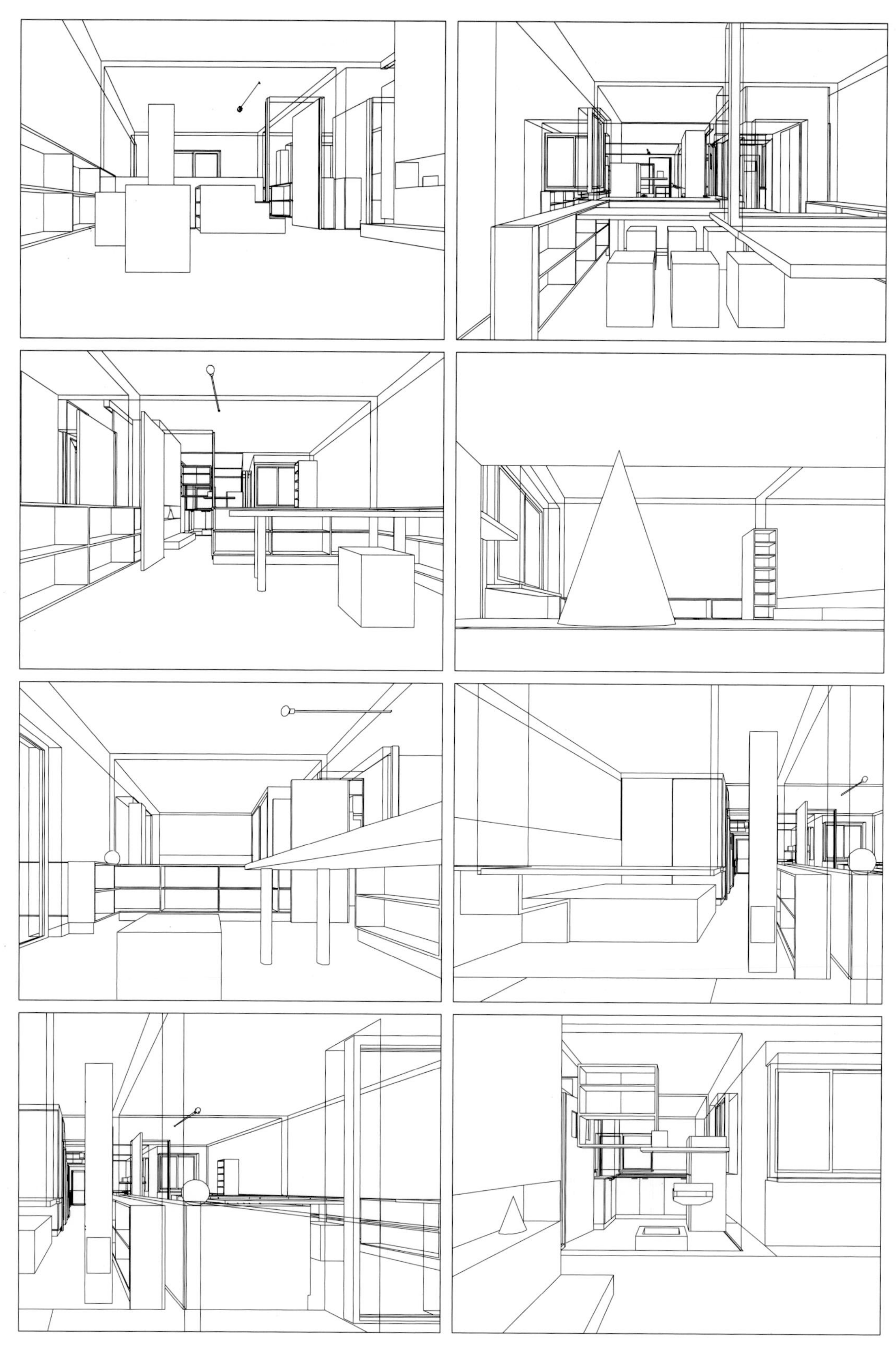

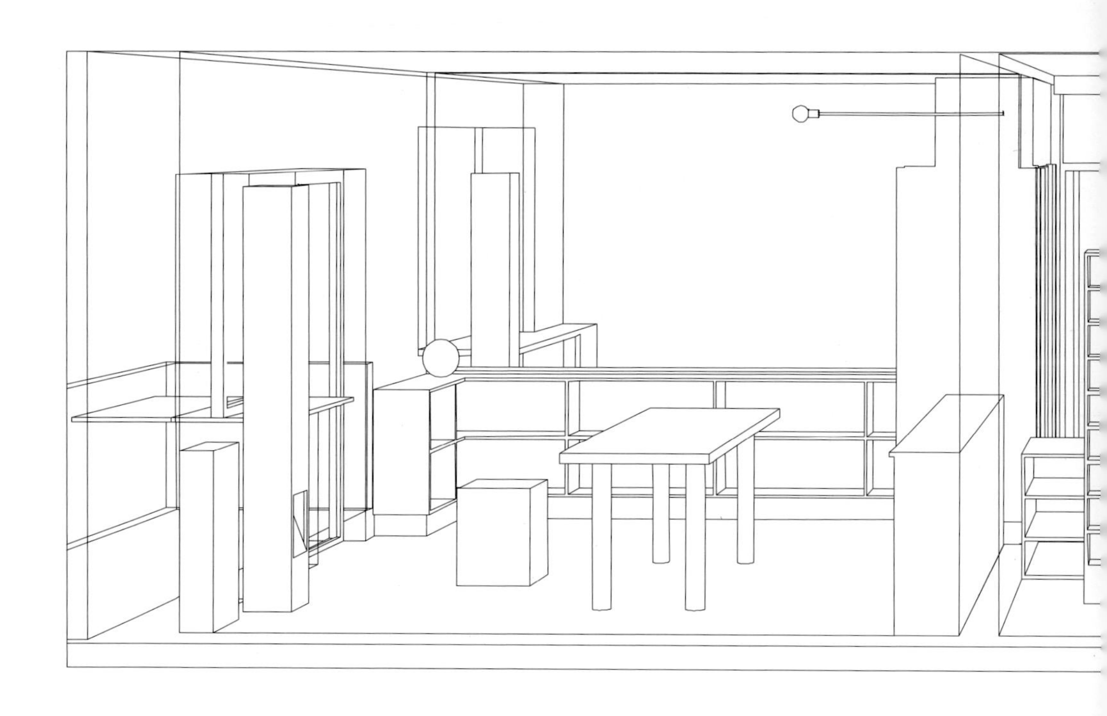

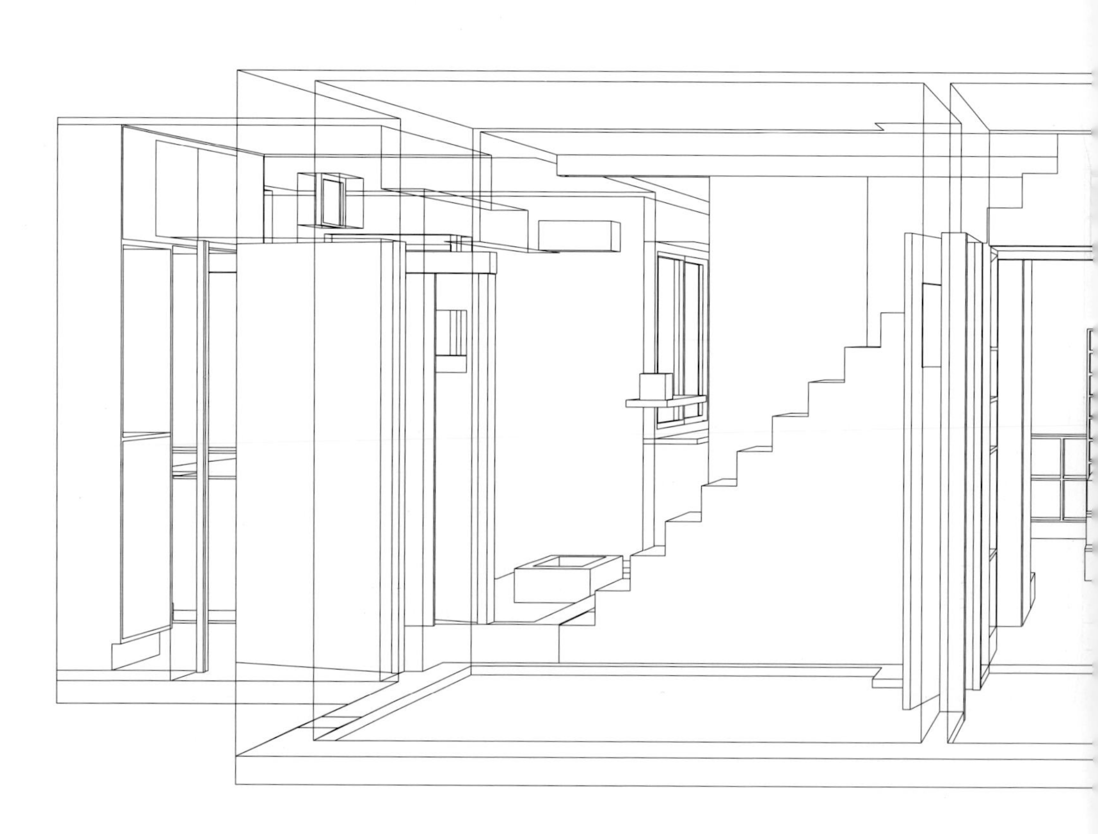

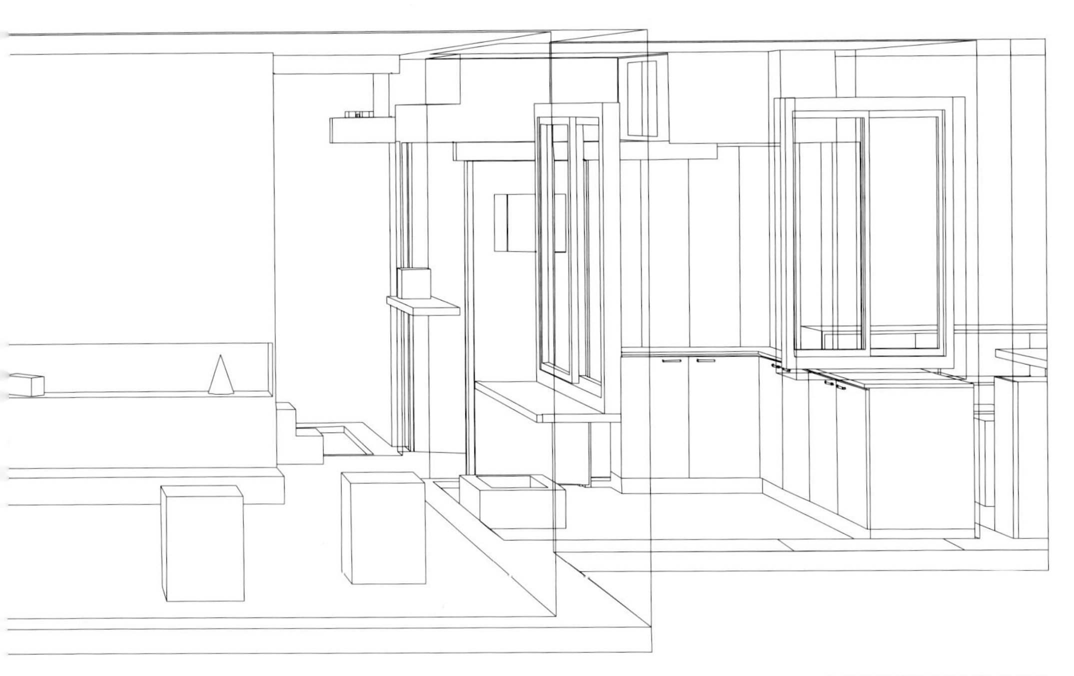

从东向西侧整体空间的关系图
The general spatial relation graphic from east side to west side

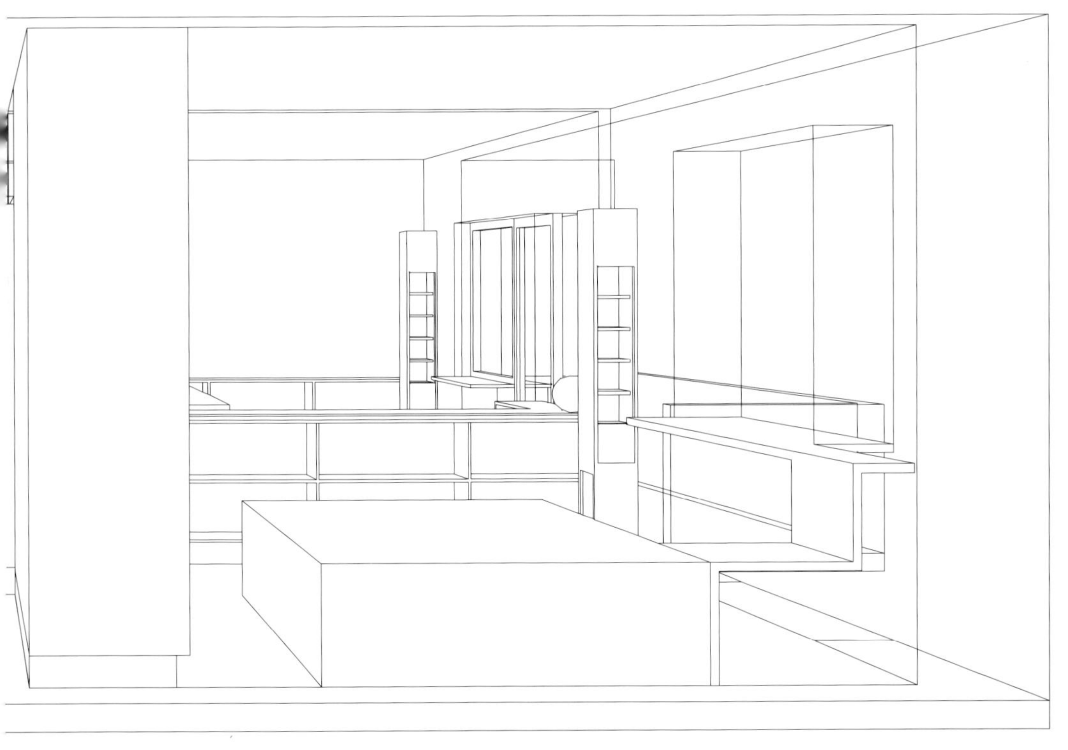

从西向东侧整体空间的关系图
The general spatial relation graphic from west side to east side

图：从极小城市“街道”看左前
单元住宅的入口
gure Left: The apartment entrance at
e front left from minimal city street view

造关系上，家与城市实际上是一致的。比如家的起居室、客厅与城市的广场是相互对应的，餐厅又与城市的饭馆相互对应，卧室与城市的旅馆相互对应，书房与城市的图书馆相互对应……而家里的过道又如同城市或聚落中的街道，起到将这些功能空间串联起来的作用。

沿着这样的思考，一个与城市空间构造相对应的“家”的风景便随之展开：城市远处若隐若现的钟楼，天空浮动着的云，长长街道两边穿插着窄窄的小巷，街道两侧并排的住宅。由建筑物所围合起的小广场中，人或是交谈，或是散步，或从广场匆匆穿过消失在街道的深处。入夜后，广场上的大屏幕上变幻的影像以及灯光下闪动的人与影。

伴随一幕幕鲜活场景地浮现，设计时城市与家的尺度在头脑中进行着频频的切换。也正是由于这些场景的尺度在观念中发生着变换，布局与景象也自然而然地附着到正在进行设计着的空间布局中。或许可姑且称之为我们所理解的经验与设计间相互糅合的过程吧。

空间与领域

人对空间及领域范围的认识是从非常具体的一个周边事物开始的，以这个对象物为参照，人的眼界慢慢并逐渐地放开想像，去认识世界，认识宇宙……所谓“一沙一世界”或许便是此意。反过来，城市里面的广场、图书馆、旅馆、餐厅等，它们之所以存在，也可由人的自身行为及自己的住宅中领域的划分来进行想像与认知。

为了能够容纳更多人，城市的广场空间变大；为了能够容纳更多人看书，城市的书房空间变大，形成图书

has the same factual logics as city from the perspective of spacial construction. For instance, living rooms and sitting rooms can be considered as squares in cities; dining rooms Is naturally compared to restaurants; bedrooms are the hotels and studies are the libraries. The aisles at home are similar to the streets in cities, which connect each part of functioning spaces.

Along with this logics, the view of homes with the corresponding frameworks of cities can be stretched out: the remote bell tower can be vaguely seen; several clouds float; alleys and streets interpenetrated with each other; there are residency buildings alined with roads. People are chatting, walking or quickly passing by the square surrounded by buildings. After the sunset, the big screen on the square sheds colorful lights on people, which leaves shadows.

Because of these vivid imaginations, the measurements of home and city are switching constantly while I was designing the plan. And also because of this, layouts and views are naturally included into my design. It could probably be called as a procedure combining experiences and design.

Space and Region

Human's cognition of space and reign begins from very specific items around. Using this as a reference, our horizon has been broadened to imagine and get to know this world, even the universe. The old sayings "a sand contains a world" probably has the same meaning. From the opposite point of view, the existence of squares, libraries, hotels

右图：从“图书馆”通往“广场”的入口
Figure Right: The entrance from the “library” to the “square”

馆；为了能够容纳更多的人去吃饭，城市的餐饮空间变大，形成餐厅、饭馆……

从这种人的行为角度来进行思考，家与城市的这种对应关系是成立的，而且从尺度上看，家就是一座最小的城市。

具体到自己的这个家，由于其实际面积只有 $60m^2$ 大小，因此设计时我就把它作为一个 $60m^2$ 的极小城市来看待。而一旦从这个角度进行切入和思考时，家里的所有家具，瞬间便成为城市当中一个又一个的建筑要素。因此对于这些家具所处的位置、摆放方式及采用的形态，也必须从城市规划的尺度和角度来对其审视和看待。因为这些家具本身的位置既相互关系，又会对未来的这个60平方米极小城市的形态产生影响。

在这个被虚拟放大到城市尺度上的室内环境中，布置和安排在其中的一切成为超大尺度的对象物。居室中安放的大柜子，是城市当中的一个大的板楼。直立着的承装CD的架子，是城市中的高塔……如此这样，整个居室空间变得缩放迷离。人行走于其中，不同角度，不同时段，唤起个人记忆当中的不同碎片……整个居室的物理空间与行走其中的人的心理空间之间，产生着呼唤和共鸣。

空间与设计

沿着上述思考，具体设计时进行了如下展开。首先在现有的起居室中隔出一条“街道”，分出街道和广场，“街道”的尺度是非常窄小的，仅有0.8m宽。在街道的另一侧，希腊米科诺斯岛中住宅将卫生间放在室外住宅楼梯下方的设计手法被直接移植到这里。有所不同的是：这里的“楼梯”并不是楼梯，而是一

and restaurants in cities is because of our imagination an understanding of residency.

In order to contain more people, the area of square is enlarged, so are libraries and restaurants. From thi perspective, home is the smallest city.

Specifically to this home, due to the limited area, considered this rented apartment as a minimal city. Once thinking in this way, all the furniture at home suddenl became the important architectures in city. The setting o furniture would have great impact on the future environmen of this $60m^2$ minimal city. Therefore, the arrangemen of furniture has to be dealt with from the aspect of city planning.

In this virtually enlarged interior environment, everything decorated or arranged becomes super-sized. A big wardrobe in the living room is compared to a huge building the up-standing CD shelf is the tower in the city. As walking in the living room, different time or different angles car bring back different pieces of memories. The physical space and mental space are corresponding to each other.

Space and Design

Along with the thoughts mentioned above, the specific design was developed as follows. A 0.8-meter-width street is set in street to separate square. On the other end of the road, the design of bathrooms in households on Mykonos island which is under the outside first floor staircase, has been implanted here. However, something different has been adapted: the "stairs" isn't real stairs, but a stair-shaped wardrobe that is against the wall. Hence the door

左上图："街道"上的"路灯"
左下图："广场"与"映像"

Figure Up Left: The "road light" on the street"
Figure Down Left: The "square" and the "image"

个楼梯状的紧挨着墙壁的柜子。居室中卫生间的门与"楼梯"下方壁柜的门被合二为一，由此卫生间自身便成功隐身。同时，向上的梯阶状的柜子一方面避免了空间中拥堵的感觉，另一方面也可以将人的视线引向空中。

在街道的另一侧有一个被称为"橱窗"的巨大的柜子，柜子的门采用推拉方式打开与关闭。"广场"和"街道"之间的关系可以随"橱窗"门的开启与闭合而改变"广场"的开敞与封闭的性质。"街道"的尺度虽然很窄，但纵深是可观的。如果加上其向南北两侧房间的"延长线"，这条街道足有10m之长。位于卧室内的"圣·几米尼亚诺的塔"是这个十米长街的端点与视觉收束点。

"广场"是家的中心场所，是聚集与交际的空间。在60m² 极小城市广场的一侧，正好是上面所描述的那个"街道"中"橱窗"的背面。这个"橱窗"的背面恰好可作为广场上"标志性建筑"的主立面。这个"标志性建筑"所呈现出的造型，是"一贫如洗"的状态。整体既不扭捏，也不作态。这个"一贫如洗"的建筑立面，时而呈现空白状态，时而又由于投射其上的影像而变得变化与斑斓。因此使得"广场"空间的动与静关系因时段而格外分明。

"广场"上还有围合的"摆放物"，有密斯"少就是多"的摩天楼状的书架、低矮舒展可供储藏和放置录像机与音响的多功能使用的"大台阶"……

因地制宜

在进行空间与设计的过程中，由于原有的空间布局

of the restroom is automatically invisible. The upward stair-shaped wardrobe not only avoids the overwhelming feeling of space, but also leads the eye sight into the air.

On the other side of the "street", there has a huge cupboard called "window", the door of this cupboard can be opened by pushing and closed by pulling. The relationship between "square" and "street" can be changed as the open and close of the "show window". Although the width of the "road" is narrow, the depth of it is visible. But if the "expansion-line" of the north and south-sided bedrooms are added, this "road" has the length of 10 meters. "The San Gimignano Tower" located in the bedroom is seen as an end of this 10 meters long street.

The "square" is the center of home, which is also the place of gathering and socializing. On the one side of this "square" is the back side of the "street" and the "show window" mentioned. The backside of this "show window" is happen to be the main vertical surface of the "landmark" in square. This "landmark" shows a status of "nothing but clearness". There is no affectation at all. This clear vertical surface only states with nothing, but it can be projected with changing images and colors. Thus the space of the square is a demonstration of a relation between quietness and motions.

There are also some devices surrounded the square, the skyscraper-shaped book shelf, the multifunctional stairs with storage space, etc...

Suit the Measures to Local Conditions

During the procedure of designing, because the original layout of this apartment cannot be revised, how to suit the

图和右图:“广场”的一角

gure Left & Right: both are one corner of the square

右图:透过“街道”的缝隙看“广场”的场景
Figure Right: The view of the 'square' through a gap from the 'street'

无法修改，如何利用现状进行因地制宜的改造成为一种新与旧的博弈。

改造前，南侧的两间居室的门都是朝起居室开启的，这两扇门的存在造成了起居室中这两扇门的周围空间无法被充分使用，而只能作为交通空间使用。同时也由于这两个开口很大，因此造成开口前面的一大部分空间成为过厅。如何控制、限定这个过厅状的交通空间，使其不被无节制地放大，同时将这个交通空间有效地利用起来是设计过程中遇到的第一个需要解决的难题。

设计时，在这两个居室门前面安放前文提到的“橱窗”——一个巨大的柜子。布置时让柜子的进深一侧作为阻隔两个居室门前交通空间蔓延的工具。通过调整宽度的设定，获得“分隔”这个动作，有效地围、堵了这两扇门所造成的空间不安定性格的无节制蔓延。限定了交通空间泛滥的同时，也使得这部分的空间性质变得暧昧，进而也将上面所提到的起居室空间中所分隔出来的“广场”空间本身进行了两种不同性格的划分，即除了主体的“广场”空间和“街道”空间外，还出现了连接这个“广场”空间和“街道”空间以及两个居室的一段很小的过渡空间。通过这个过渡空间，使广场在侧面形成了一个有层次且有深度的空间感觉。

“广场”背后的“街道”贯穿住宅的南北两端，前面曾提到这个街道从南侧的卧室到北侧的餐厅总长约有10m。这条10m的“街道”是整个住宅中最长的进深。改造前的户型中，这个进深被两个房间切分开，使得住宅空间中最“奢侈”的一部分被忽略了。在新的设计时，有意识地将这10m的延展部分进行一览无余地释放，无

measures to the current conditions had become a zero-sur game of the new and the old.

Before the renovation, the two bedrooms on the sout side are opened towards the living room, which made th space between these two doors become nothing but fo commuting. Meanwhile, because the size of those tw doors are relatively big, the opening space can only b used as part of the living room for passing by. Hence, hov to effectively utilize this space had become the first probler that needs to be solved.

When conducting the design, I set the "show window"—th huge wardrobe as an expansion to separate the openin space. Through adjusting the width, I was taking control o the unstable characteristics of the space created by thos two doors. At the same time of limiting the commutin space, this part of the area had become vague, furthermore the "square" could be apart my different characteristics, tha is, besides the space of "square" and "street", there is als a transitioning part connecting them, which cultivates th feelings of depth.

The "street" on the back side of the "square" runs throug the north and the south end of this apartment, as it i mentioned above, it could reach to the length about 1(meters which is the longest spatial length in this condo Previously, this length was separated by two bedrooms which made the most spatially luxury part ignored. In my new design, I intentionally release the potential of this 10-meter expansion to the fullest. When standing from one end of the "street" looking towards the other end, you wil find it is a space with multiple visual depths, which als

图：“卧室”中作为灯具而存在“圣几米尼亚诺的塔”

gure Left: The "San Gimignano wer" as a lamp in bedroom

形中将住宅室内的空间进行最大可能性地展示。当我们站在“街道”的一端向另一端望去时，这条“街道”中会形成一个极富景深的视觉效果，通透的视线中还展示着层叠的空间关系与效果。

空间与分割

分割是空间运作的重要手段，是产生空间韵律与节奏的关键。相同面积的空间，由于分割方式的不同，空间感以及具体的使用方式都会有所不同。因此在近乎相同面积的前提下，空间感的大小问题与空间的切分方式密切相关。还是以前面所谈到的起居室为例来谈这个空间的分割问题。

在构思这个 60m^2 极小城市的阶段，已经决定要在客厅中安放一个“大柜子”——“橱窗”，这个“大柜子”——“橱窗”的安放，初始仅仅是考虑要将居室与卫生间隔开这样非常简单的一个初衷。然而就在图纸中展示“隔开”行动的瞬间，起居室空间便分隔成两部分。由于居室的面积是一定的，隔开切分时，被分开的两个空间，彼此之间的具体尺寸的确定问题就变得非常突出。

为了让“广场”足够大，相应地“街道”就要足够地窄。所谓“疏可走马、密不容针” 或许便是此意。而关于住宅中的走道宽度的设置问题，按规范和一般概念层面来理解，最低也要设计成 1.1 米宽左右。为了让“广场”感觉“疏可走马”般宽阔，60m^2 极小城市中的“街道”便“谨小慎微”地确定为 0.8m 宽，同时将“密不容针”“街道”的另外一侧楼梯的宽度设定为 0.45m 宽。

lets you see through the relations and effects of spaces overlaid.

Space and Separation

Separation is a significant method of operating space, which is also the key element of creating rhyme and rhythm of space. Spaces having same areas, if apart not in the same ways, will bring very different feelings and specific usage to people living in there. We are going to use living room mentioned above as an example to illustrate the spatial separation.

During the conceiving phrase of this 60m^2 minimal city, I'd already decided to arrange a big wardrobe—the "show window" but merely to separate living room and restroom. However, at the moment of seeing the "separation" on the blueprint, I found the space of living room being automatically separated. The area of living room is settled, so the specific sizes of the two spaces have come to the center of the stage.

In order to guarantee enough space for the living room, the space left for the "street" has to be relatively small. As an old Chinese saying "If the space is ought to be left spatial, it should leave enough room for riding horses; if the space is ought to be left narrow, then wind shall not be getting through", the very basic concept of setting aisle should have at least 1.1 meter width in order to let people feel enough space. Nonetheless, I set the width of the aisle of this minimal city as 0.8 meter and the width of the stairs at the other side of the “street” as 0.45 meter.

About the width of the stair, basically, a standard stair would

右图："路灯"与"街道"中的"楼梯"相互叠加

Figure Right: The overlap of the "road light" and the stairs in the "street"

关于楼梯的宽度问题，一般地，作为一个标准的楼梯至少要有 0.8m 或 0.9m 以上的宽度，但是这个狭窄的楼梯，由于与以往经验的比对，客观上错乱了观者的尺度感觉。确切地说，当对这种观念上既存楼梯的尺度进行现实层面上的压缩之后，楼梯就会在人的观念上与现实中的楼梯产生一种错觉。当在看到这里所摆放的这个变异了尺度的狭窄的楼梯，似乎还会感觉这个楼梯就是平时走惯了的那个正常的楼梯的尺度。而一旦用这个楼梯所带来的感官上的尺度错觉再去衡量这个住宅室内的所有其他空间时，住宅的空间便会在观念中被放大。

经过实际使用，"谨小慎微"地确定为 0.8m 宽的"街道"没有任何的不便。据此，走道的宽度能否在此基础上再窄些或许成为下一个项目的挑战。

这个"大柜子"——"橱窗"所形成的"物件"在原本封闭的起居室空间中划分出一个大空间——"广场"和一个小空间——"街道"的同时，还在"大柜子"——"橱窗"的两侧分别留出两个豁口，目的是让所隔出的"广场"与"街道"之间形成一个"回路"，让这两个空间得以循环。当人从不同方向围绕其行走时，体验到空间的流动性的同时，还获得一种"迷路"的感觉，从而扩大人的空间感受。

空间与体验

空间设计时对于未来空间体验的"设定"是件重要的事情。设计师需要预知观者在空间体验中可能获得的某种感受的状态。

从住宅的入口进入这个 60m² 的极小城市中，迎面

have at least 0.8 to 0.9 meter of width. But if we apply it t this case, it will give the audience a contradictory feelin instead. To be more specific, when the size of the stair i shrunk to fit into a narrow space, this create an illusion b tricking people to believe that is the logical size. Meanwhile using this stair as the reference to measure other area i this space, people will conceptually amplify the area of thi apartment.

In practical, the carefully set size of 0.8 meter is proved t be convenient as the normal size staircase. Based on this there's another challenge raised, that is, whether the aisl could be shrunk a little bit more. This big wardrobe—"th show window" draws a line between one bigger spac the "square" and one smaller space the "street". It als creates a loop between these two spaces and leads then to connect with each other. People could come and g into these two spaces from different directions, thus it ca enlarge people's sensation of the overall space.

Space and Experience

It is rather important to pre-set the future spatial experience Designer need to predict the potential living experience.

From the entrance of this minimal city, eyesight is blocke by the closed door of kitchen and walls, turn right, come to the space of the "street", view could be extended i this narrow but long area; once passing by the narrow commuting space, you could see a somewhat spaciou "square".

These extension, coverage and blockage could lead t psychological change and sparks of the audience, which

上图："广场"一角
下图：从"广场"看"街道"及往卫生间的入口

Figure Up Left: One corner of the "square"
Figure Down Left: The view from the "square" to the "street" and the entrance of bathroom

的视线被封闭的厨房门和墙体所阻挡，向右转身，进入"街道"空间，视线在这个狭长的空间中得到延伸。透过狭窄的过渡空间，可以看到一个宽敞的小广场……如此这种使观者在视线上出现的延伸、遮挡、阻断可以让人在心理感受上产生变化和碰撞，这也是设计所考虑的要素。对于在这种思考下所形成的建筑平面图，看上去似乎呈现的是一个图案，但实际上建筑平面中所表示出的布局与分隔，直接影响未来在这个建筑空间中生活与活动的人的视线以及视线所产生的人的心理感受之间相互的碰撞关系。

作为壁柜而存在的楼梯位于"街道"的中途，由于这个向上的楼梯的存在，人的视线在这个地方又多出另一个向上方向的选择。这个视线方向的变化，在人的心理上形成一种复杂性的期待。

极小城市中的"街道"空间在现实中是非常暗的，尤其是当"大柜子"两侧的推拉门关上时。只有位于南侧的卧室和北侧的餐厅内的自然采光可以照亮"街道"的两个端头。当人的视线在这条街道上扫过的时候，会经历一个从明亮到昏暗，再回到明亮的过程。而这个空间中视觉明暗的对比又会因为时间和季节的变化而产生不同。

当"街道"中"大柜子"——"橱窗"两侧的推拉门打开时，人站在这条"街道"上，会发现"街道"并不是一个简单的空间，而是时而封闭、狭窄，时而开敞、宽松的状态。这种充满着韵律的变化，会使身处其中的人感觉到这条"街道"的丰富，在人的心理感知当中产生出多种的变化。

实际上，"橱窗"在整个"广场"和"街道"的空间层次的塑造上起着非常大的作用。在设计"广场"

is also the essential element that needs to be considered while designing. Therefore, the designing under these considerations seems to be just a pattern, but in fact they have direct impacts on future living experience in this space, and also the feelings attached to them.

The cupboard shown as the "stair" is located in the middle of the "street". Because of the existence of this upward sloping "stair", there's an alternative for peoples' eyesight, which forms an expectation of complexity.

In reality, the "street" of this minimal city is very dim, especially when the slide door of big wardrobe is closed. Only the natural light from the bedroom on the south and the kitchen on the north could lighten up the two ends of this "street". When people set their eyes on this "street", they will experience a process of brightness to dimness and back to brightness. And these changes of light will be altered along with time and seasons.

When the big wardrobe—the "show window" is opened, the audience will find that the "street" is a space with somewhat complex status: sometimes closed and narrow, sometimes open and broad. With the plenty of varieties, people who set their feet on this "street" will sense the diversity of this space.

In fact, the "show window" plays a huge role in constructing the spatial layers of the "square" and the "street". I was hoping the "square" could cultivate a spatially progressive view by emphasizing the back and forth allocation of items.

There's one thing worth mentioning, that is the "road light" of the "street". It is located in the middle of the "show window" and the "stair", which effectively adjust the overlay

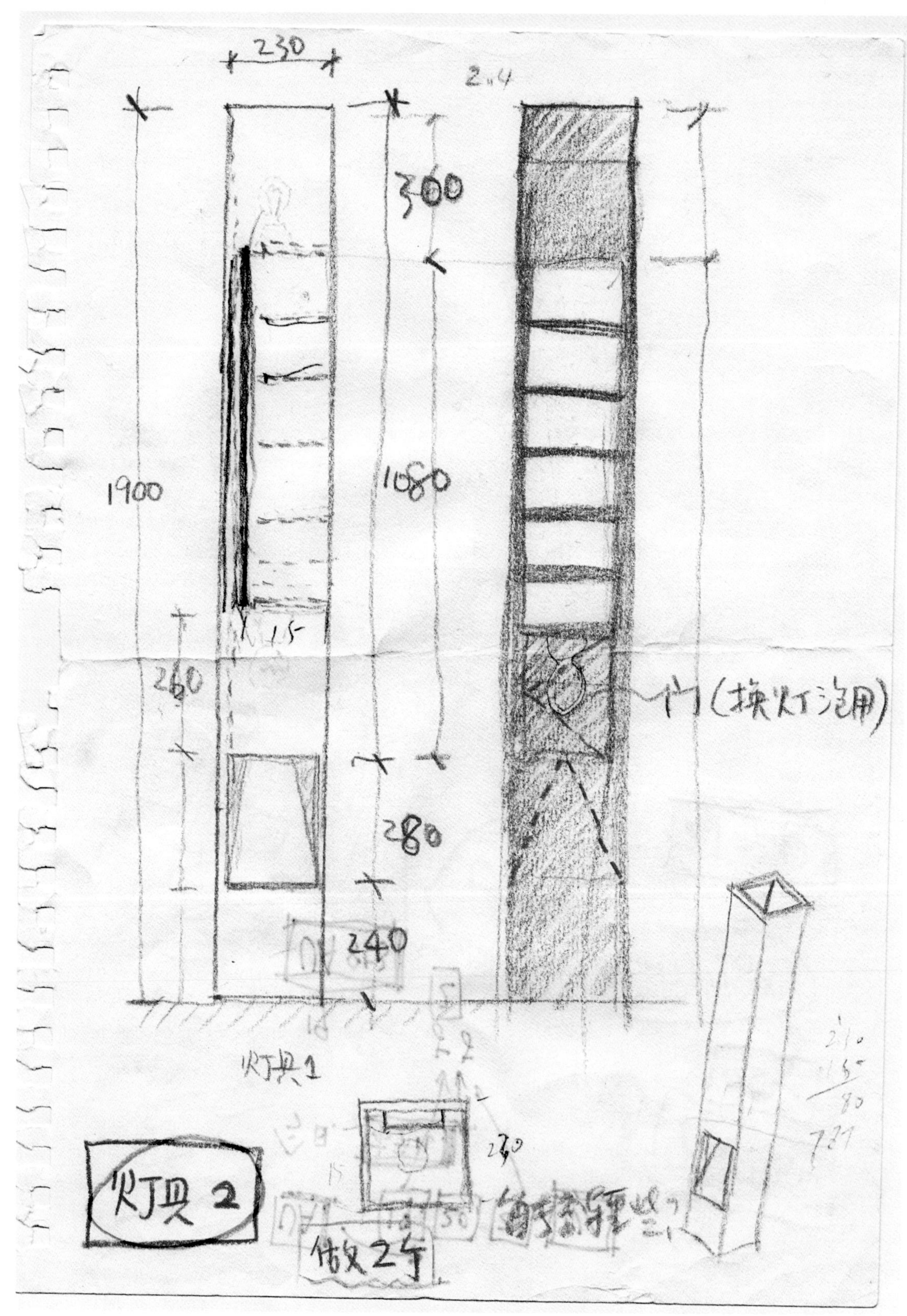

“塔”的设计草图 The design draft of the "tower"

“卧室”的设计草图
The design draft of "bedroom"

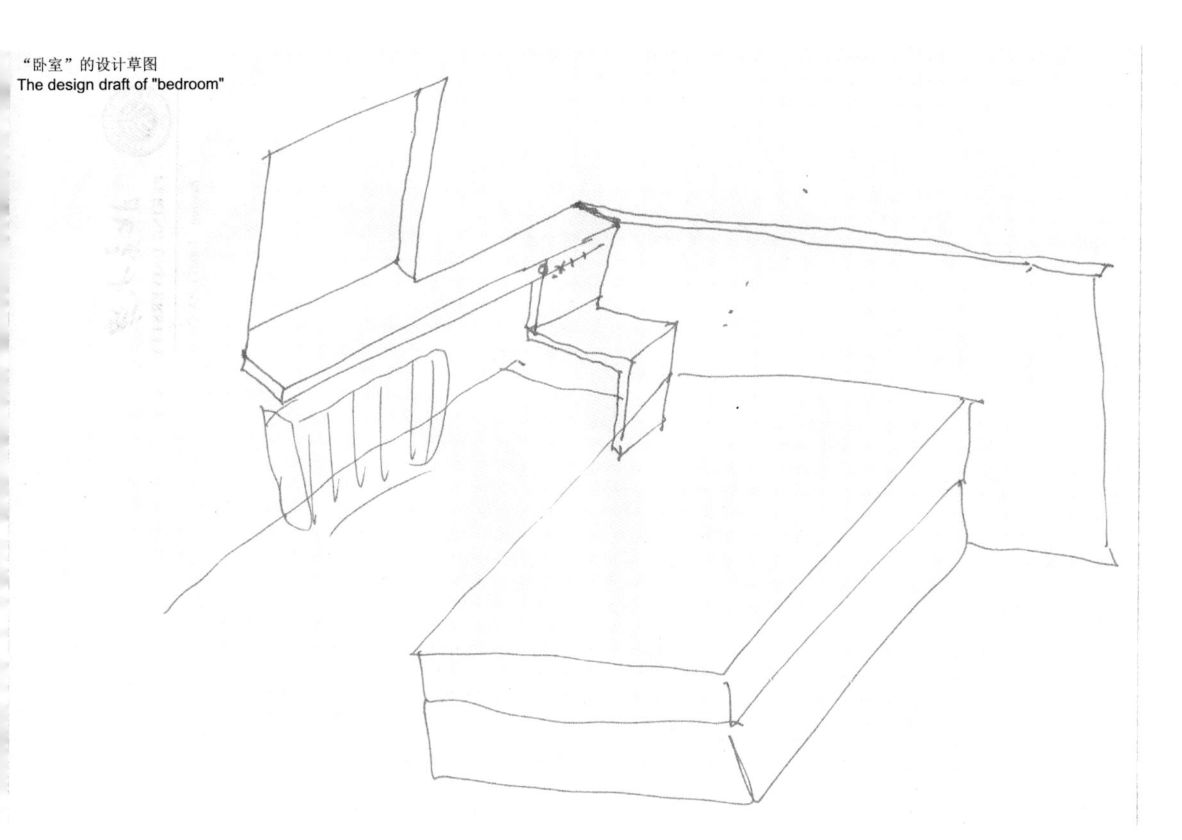

“广场”上的“储藏”设计草图
The design draft of the "storage" at the "square"

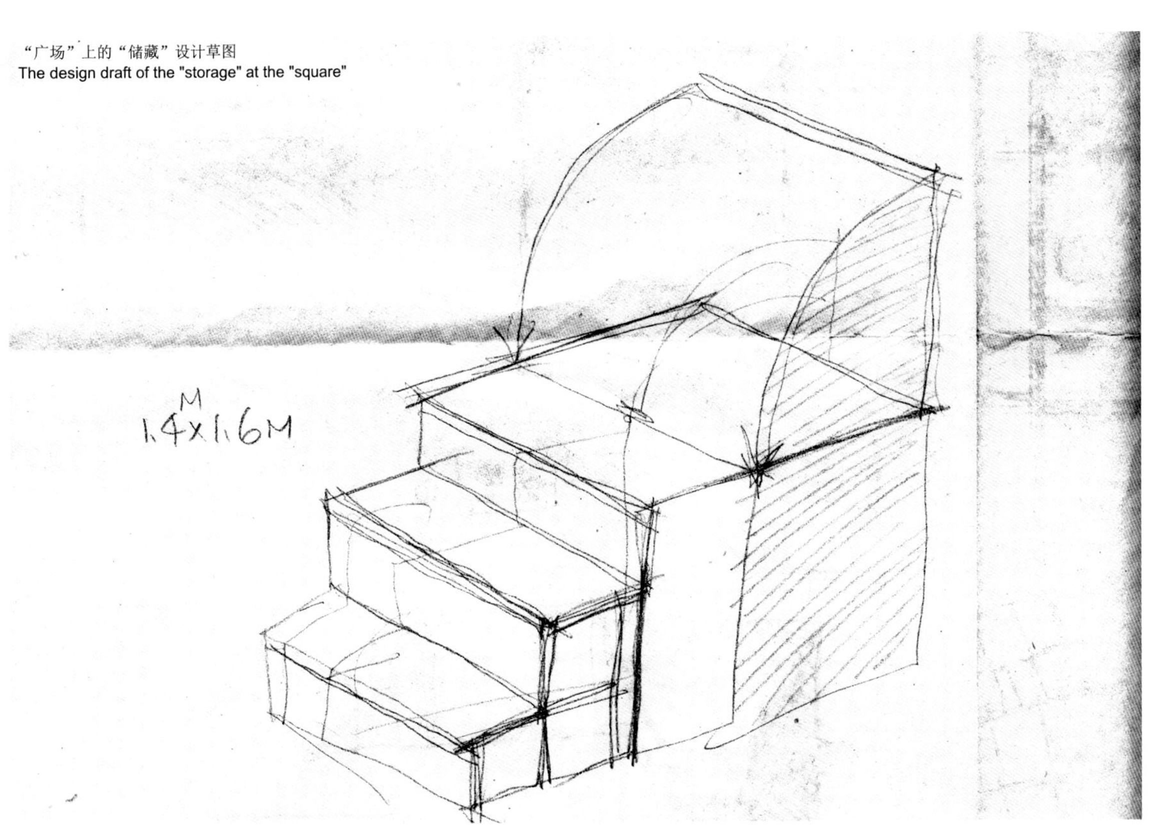

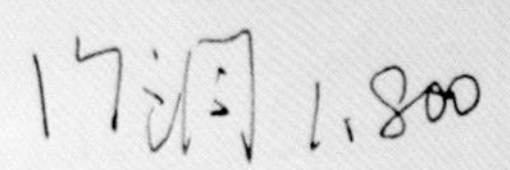

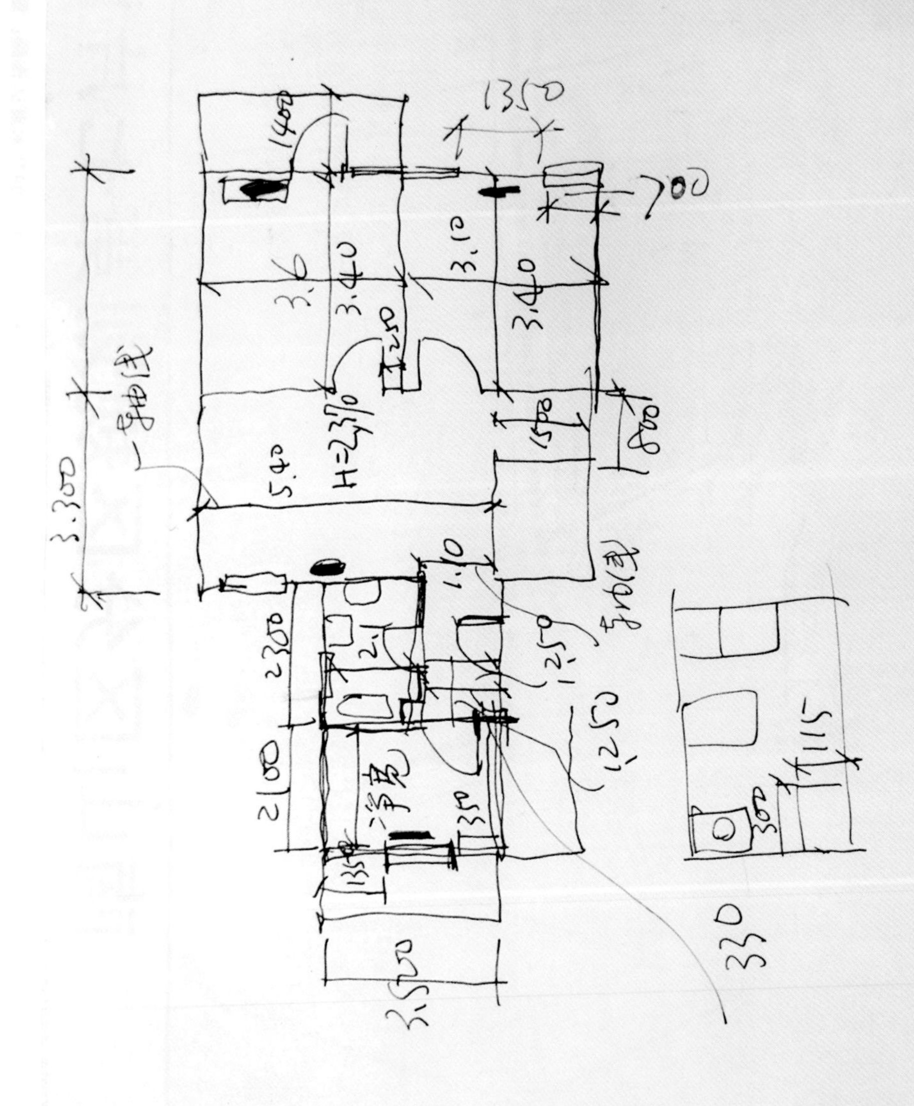

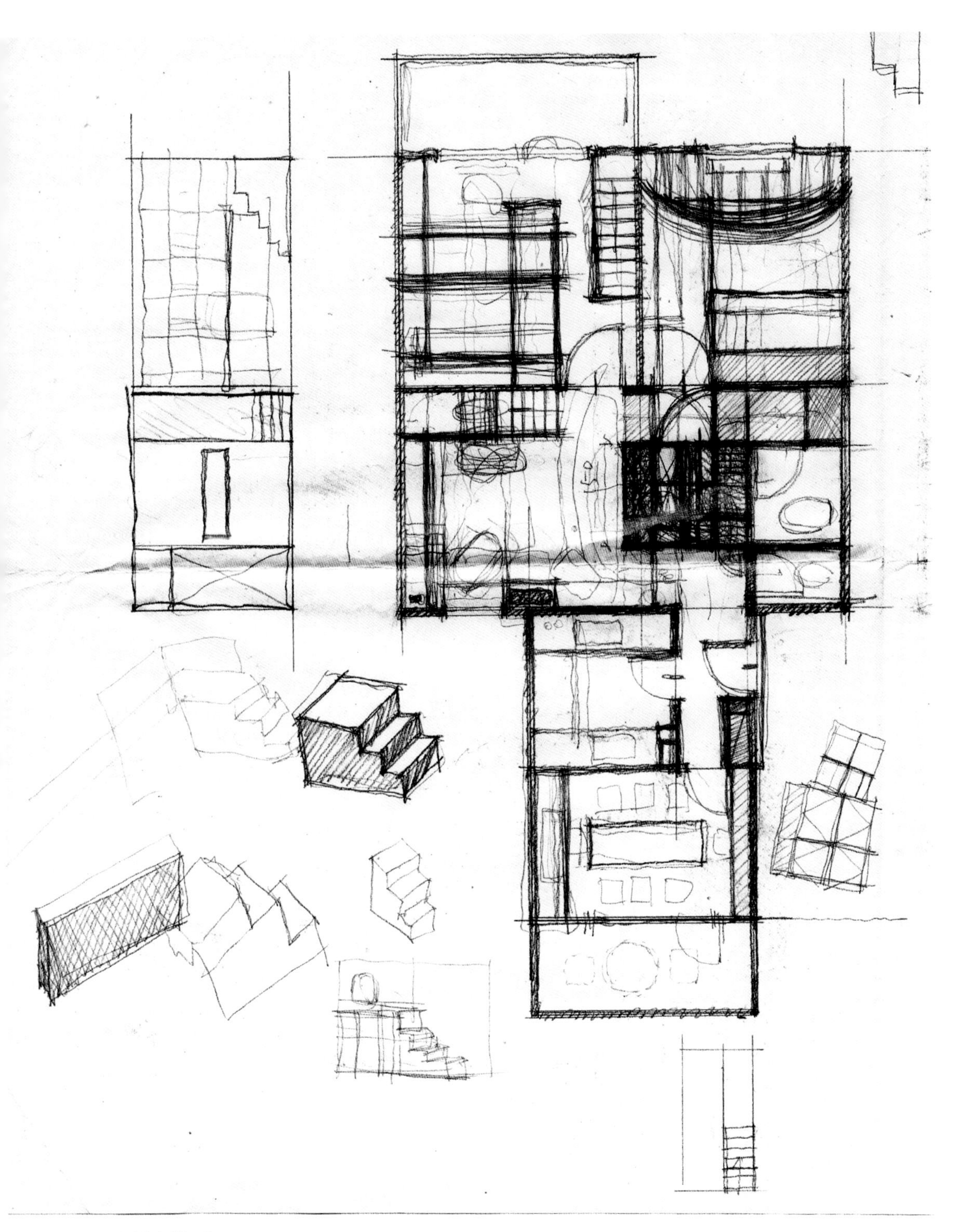

左图:改造前针对极小城市当时的现状所进行的测绘图
右图:极小城市的设计草图
Figure Left: The mapping diagram of the original condition of the minimal city
Figure Right: The design draft of the minimal city

左图：“广场”鸟瞰
右图：“卧室”一侧的大壁柜
Figure Left: The aerial view of the "square"
Figure Right: The closet of bedroom

左图：卧室的一角
Figure Left: One corner of bedroom

时，希冀“广场”能营造出一种空间景象层层叠加、形成逐级递进的层次感，因此设计时尽可能强调物体与物体之间的正对的层次性表达，强调空间的平面性，尽可能更容易地形成叠加的效果，通过强调物件前后、进深、凹凸等增加视觉上的层次与丰富感。

在这个“广场”空间中，有一盏“街道”上的“路灯”是值得一提的。这盏灯的所在位置，成为“广场”上的“大柜子”——“橱窗”和远处的“楼梯”中间的层面，从而有效地调和了远近两个层次在叠加后的平面感。

视线在空间中的延展与伸缩，以及这种延展与伸缩对于空间感受的影响也是改造过程中重点思考的内容。因此，设计时有意识地对视线的走向进行控制。具体地，当视线透过面前的一个“隔”物，看到它延伸处的一扇窗，再透过这扇窗看出去的这个过程中，视线能够获得无限地延展（如沿着住宅南北两侧房间的“延长线”的超过10m之长的街道望到窗外的情形）。而当视线稍微偏转，立即因被其他物体遮挡而反弹……当沿着“街道”在室内空间中游走时，人的视线在“长”与“短”之间不断变换，视线在获得出口时得以延伸，在遇到障碍物时被反弹回来的瞬间，不同的空间在心理感受中得到不同的比较与测量。

设计与“引用”

设计中，“引用”概念的本身有两种含义；一是在现场对于一些部品进行物质层面的引用，另外是对曾经历和看到的一些场景与做法的引用。60m^{2}极小城市的“引用”是在这个理解层面并结合具体的现场来

of remote items and the ones nearby.

The extension and shrinkage of eye-sight and their effect on space experience are also the significant part of designing the plan. Therefore, I intentionally control the direction of eyesight. Specifically, when eyesight falls through an "obstacle" and reaches to a window on its expansion, then to the outside of the window, the view could be unlimitedly extended. However, if the eyesight goes to a slightly different direction, it will be blocked immediately. Hence, the psychological changes could be measured and experienced along with different eyesight.

Design and "Quotation"

There are two sets of concepts of "quotation" in designing, one is to utilize similar materials on site; the other one is to use the elements of what have already been seen and experienced. The "quotation" used in this minimal city is to use the latter concept and combine the settings on site.

The first to be mentioned is the kitchen. In the original design, the entrance door of the apartment is right facing the door of kitchen, which has glass window on it. That is to say that, when you step into this apartment, the first thing get into your eye is the transparent door of the kitchen, and the messy in the kitchen. Because of the structure, the area of kitchen can not be changed. Therefore, the key point is to change the door. The bedroom on the north side had been changed to the dining room, so the door could be left out. So i use that door and "quote" it to kitchen in order to replace the glass door. Under this setting, if the door of the kitchen is closed, a full wall is formed instead of the kitchen,

右上图：“图书馆”的灯具
右下图：围合“广场”的“大板楼”（书柜）鸟瞰

Figure Up Right: Lamp in "library"
Figure Down Right: The aerial view of bookshelf at the "square"

展开的。

首先要提的是厨房。在原设计中，打开住宅的户门，正对着的就是厨房的门，而且是一上部有玻璃窗的门。就是说，一进门第一眼所看见的就是厨房的门和透过玻璃窗所看到的厨房内部的凌乱。由于结构的原因，厨房空间无法进行位置变换，因此改造设计的重点便放在了“门”的处理上。由于北侧的居室被改成了餐厅，所以这个餐厅空间的门就不需要了，把它拆下来，直接“引用”到厨房，安装上轨道，设计成推拉门，并将原来的那扇带有玻璃窗的门替换下来。这样做的目的是，当把厨房门关上时，入口处形成的是一个完整的墙面。当从户门进入到室内的时候，眼前呈现的是一个立体的“白中之白”，几乎猜想不出其后就是厨房。

对于“门”的关注，同样地也作用在卫生间的改造设计上。开篇时谈到，这套住宅最大的症结就是卫生间的设计问题。因为卫生间的门是朝向起居室开启的，所以这个门如果说通过改造设计仍然还能让人感觉到是卫生间的门，看到这扇门就会立刻意识到它是卫生间。为了通过改造设计能够“隐蔽”卫生间的存在，这扇门的设计成为需要解决的问题中的一大关键。

前面谈到，在卫生间的改造处理上，引用了曾在聚落调查中看到的希腊的米克诺斯岛上的场景。在米克诺斯岛上，住宅的卫生间放在室外从一层通往二层的楼梯的下面，极具隐藏特征。受到这种处理手法的启发，设计时在“街道”层面的那个壁柜楼梯的下面增加了一道门，作为卫生间前室的门。这样，卫生间的门就变成了楼梯下面的一个柜门，而且这个柜门是与这个楼梯的下面壁柜的柜门相一致并加以混淆的。当这扇

so you will hardly recognize its existence.

The attention on door has a simultaneously effect on the re-modification design of the bathroom. As it was mentioned before, the biggest problem of this apartment is the bathroom design. Due to the bathroom door faced the living room, it's supposed to be recognized as the bathroom door after the modification. However, the existence of the bathroom will also somehow be “hided”. The design of this door becomes crucial.

As mentioned the bathroom re-modificaiton design is quoted the scene on the Mykonos island in Greece from the settlement investigation. On Mykonos island, bathrooms in households are designed to be under the outside first floor staircase as to show its hidden feature. This feature inspires my design of the bathroom door. I added a door under the staircase closet at the "street" as the bathroom door. So the bathroom door becomes a closet door under the staircase and the style of this door stays consistent and looks confused with the real closet door. When people enter this closet door like bathroom door, it's the laundry room. On the right side of the laundry room there will be the bathroom entrance. Through all these effort, the bathroom is completely hidden.

About Color

The usage of color elements in this re-modification is very cautious. The reason of using white as the basic color for the minimal city is that a fully white color painted room helps to cause a sense of ambiguous space and to easily blur the space boundaries.

左图：“街道”上空的“路灯”和在“楼梯”左侧还可以看到楼梯下方的柜子隔板

Figure Left: The "road light" above the "street". A view of the shelf inside the stairs

看上去像壁柜的柜门打开后，门后是放洗衣机的房间。进入这个房间，通过其右侧的另外一道门，才可进入卫生间。经过这样一番“折腾”，卫生间被彻底地隐藏掉了。

关于色彩

色彩要素的使用，在这个改造的过程中是极其慎重的。最终将白色作为极小城市的色彩是由于：一个房间，当整体色彩一旦采用白色后，空间的整体就会给人造成一种迷茫感，同时也会极易失去空间边界的界定。这是因为假如一个房间中的墙面的材料采用某种色彩，或采用某种肌理的材料，整个墙面会立即变得真实与具象，同时人对于这个墙面本身的所有的判断，也便变成了一个非常具象和具体的判断。

如果这个房间的墙面是白颜色的，特别是在光影变化明显时，极易造成人眼无法正确地判断墙面和人之间的距离的状况。这就好像给白色的房间拍照，有时相机的焦距很难对准。有时面对白色的物体，自动聚焦相机要想对准其焦距，镜头就会不停地前后移动、来回寻找，去测量、去判断，试图找到距离关系。实际上面对白色，人的眼睛也是如此。当你放眼望去，眼前看不到“对象物”。因此你的眼睛就需要不断地调节，不断地寻着“对象物”来进行不断地甄别和判断。这个过程如同相机的镜头一样，通过不停地来回测量来判断空间的大小、距离和位置。正是由于白色的这一特性，当人身处白色房间中试图去判断具体位置，而却又无法准确判断的时候，就会产生一种距离感上的偏差和错觉，使人无法感知到这个空间是一个面积很小的房间。这也是设计希望达到

This is because when one side of the wall in a room is painted in one type of color or using a certain type of texture, the whole wall would turn to be more realistic and concrete. Meantime, people's judgements against the wall itself becomes realistic and concrete as well.

Our human eyes easily give false judgements for the distance between the wall and people when all walls are painted in white, especially with strong light and shadow changes. This is reflected that it is hard to accurately focus when people shoot pictures for a fully white room. Sometimes for shooting white objects, the automatic focusing camera lens would stretch from forward to backward to search, to measure, and to estimate the distance relation in order to focus. In front of white, people's eyes do the same. It's hard for people catch concrete objects right away because your eyes are constantly adjusting in order to find signs and distinguish the concrete objects. This process is similar to the camera lens. Due to this special feature of the white color, people would get a sense of distance deviation and illusion when they are in a white room and hardly be able to accurately make judgements. So that makes people fail to notice that it is a small room. This is the exact result was expected from the design, to enlarge the sense of the space of a small room.

About Dimension

There are dimensions existing in all creatures in nature, like in human bodies, in plants, or in animals, etc. Every object itself naturally comes with its logic and rhythm. Basically two types of dimensions are related to architectural space.

左图：“广场”中的“标
志建筑”
右图：从“街道”的下方
对空中“路灯”的仰视

Figure Left: The
"landmark" at the "squa
Figure Right: A low angl
view of the "road light" o
the "street"

右图：从“广场”透过“街道”看“卧室”

Figure Right: A view of bedroom through the "street" from the "square"

和产生的效果，目的还是希望让小的房间给人以扩大的空间感受。

关于尺度

自然界中的万物都是有尺度的，比如人体、植物、动物等。每一种物体本身蕴藏有一种天然的合理性和韵律感于自身。而与建筑空间相关的尺度基本上有两种，一种是源于植物的自然尺度，一种是人体的尺度。

自然尺度往往是隐藏在结果之中的。聚落中的那些传统的木结构房子，往往看上去都很舒服。那是因为这些房子本身都有着令人舒适的尺度和比例关系。而之所以能够如此，是因为木结构的房子在建造过程中，房子的尺度受到木材原有尺寸的限制，房子要根据木材的尺寸来进行建造。比如房子的高度是由天然木材的长度决定的，它一定要取木材所能达到的一个极限值，并将其作为房子中最高的柱子之用。一般情况下，当大树干用于房子的主体部分后，那些被剃、砍下来的树木的枝杈等部分则会作为地面或者屋顶的檩条，甚至树叶、茅草都会被利用作为屋顶的面材。因此不难发现，这种建造形式，是一种自然材料以及尺度的重新构建，是通过人的这样一种建造行为，使自然材料的比例关系转换到房子当中。又由于所有的建造材料来自于自然，因此这个具有自然材料比例关系的房子必然就会与自然环境相协调。不仅美，而且看上去舒服。这是因为这种建造的过程不过就是在一个相同比例关系的环境里面，用另外一种方式把这个原有存在的比例关系重新展现出来，其结果自然会有一种相互和谐的关系体现出来。

与这种源于植物的自然尺度的建造的同时，世界上

one is from plants in nature. Another is from human body.

The dimensions in nature are usually hidden in its results. For instance, those traditional wooden structure houses in settlements look cozy due to their perfect sizes and scaling relations. These perfect sizes and scaling relations are caused by the limitation of original wooden materials during the building process. All the houses need to be built according to the original size of wooden materials. For example, the length of the wood piece decides the height of the house and the longest or the tallest wood piece would be used as one of the pillars in the house. In normal conditions, tree trunks are used as building the main part of the house and the rest of the trees, like its branches, are used for building the ground or the roof, even leaves and couch grass are used for the roof. So it's not hard to understand that this architectural formation is based on the restructuring of natural materials and its dimensions and the transformation process from materials' natural scaling relations to a house through human being's construction behaviors. Furthermore, since all construction materials coming from our nature, the whole house would look not only comfortable but harmonious with its environment. As well as the whole construction process is to show the original scaling relations in a different formation, so that the result of the construction, the house, would be naturally harmonic with its surroundings.

Other than constructions originated from natural plants' dimensions, there is another way of showing dimensions in the world. Take Rammed Earth Structure as an example. The key problem for Rammed Earth Structure is the

左图:广场一侧的“摆放”要素
右图:从卧室向北侧看极小城市的“街道”

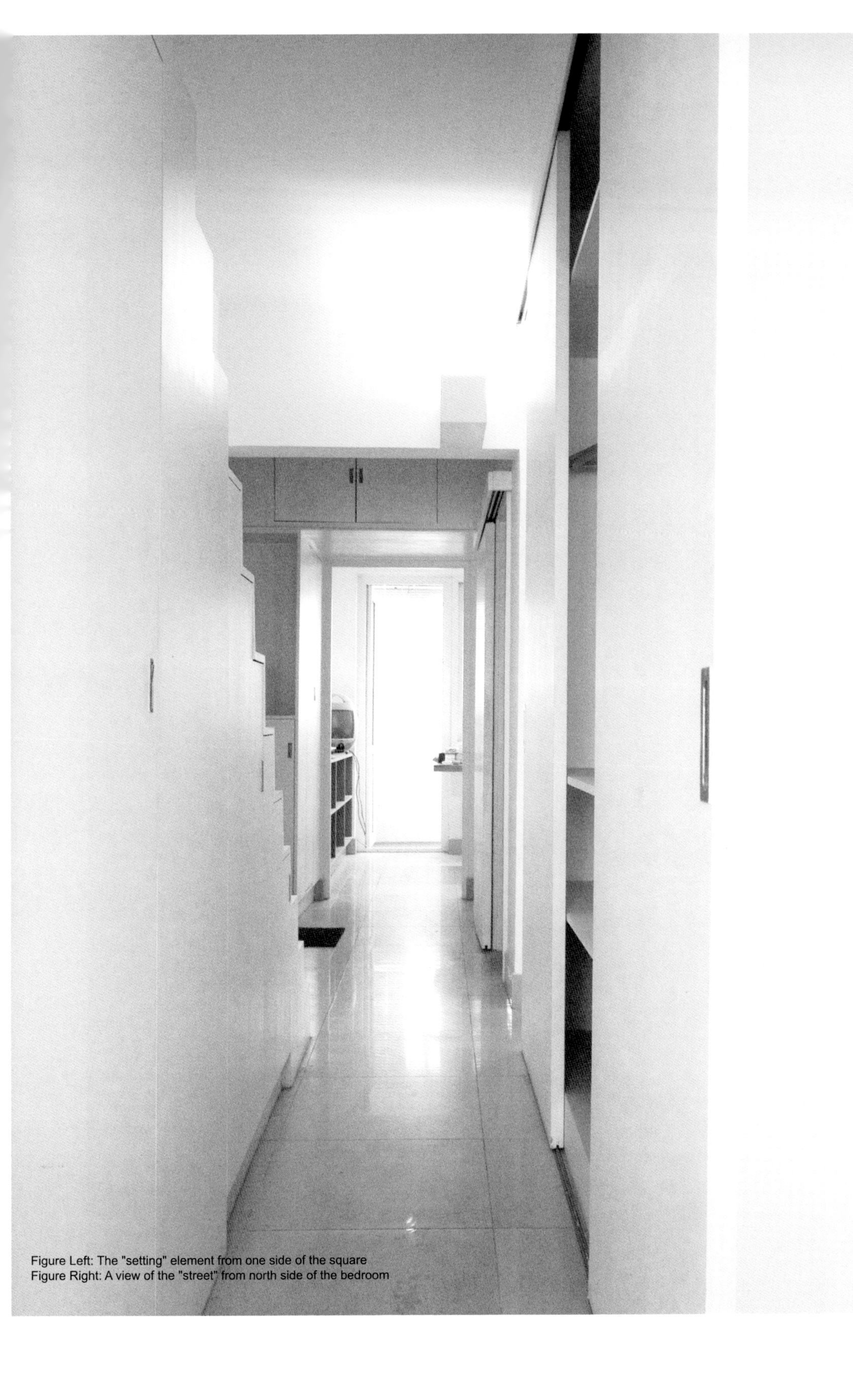

Figure Left: The "setting" element from one side of the square
Figure Right: A view of the "street" from north side of the bedroom

左图和右图:生活中的“广场”一角

Figure Left & Right: both are one corner of the "square" in life

左图:入口侧面的柜子
右图:由“街道”向北侧所看到的风景
Figure Left: The cabinet near the entrance
Figure Right: The view from north side of the "street"

图和右图：作者在极小城市中生
的场景

gure Left & Right: both are the
thor in the minimal city

右图：耸立在“广场”中的“密斯的高层建筑
Figure Right: 'Tower' at the 'squa

还存在另外一种与其完全不同的展现尺度的方式，比如夯土建筑的建造。夯土建筑在建造时所面临的问题是：它所使用的建筑材料是松散的沙土。它的建造，无法在协调感上有所依据，其尺度感如何表现？尺度如何判断？摩洛哥的房屋建造过程透露了建造者们“判断”的秘密。在那里，房子的高度、墙的厚度以及门的高度、窗户框子的宽度等，都是当地盖房子的人根据自己的身体尺度来定的。一般情况下他会根据自己手掌的宽度来决定窗框的宽度，根据自己的肘臂来规定房子墙体的厚度，然后他又根据自己的身高来规定房子的高度，因此在那里盖房子的时候，人体的尺度就会通过一系列的测量判断，于不自觉中将人体的尺度转换到所建造的对象物之中。这种由人体尺度转换而来的建筑，当其他的人走进其中或去观测它的时候，一定会从这个建筑当中感知到人体尺度固有的那种韵律感和节奏感，同时这一切会在这个建筑里面有所展示和体现。

我们之所以也会感觉夯土建筑很舒服、很美，正是因为夯土建筑的建造行为是将人体的尺度与和谐的比例关系转换到房屋当中。同时夯土建筑所使用的建筑材料又都是当地的沙石黏土，所以人们非常形象地形容：夯土建筑是从地面上生长出来的建筑。只不过这个生长出来的建筑是按照人体的尺度被重新构建了尺度关系的对象物。这样的人与尺度的关系，在 $60m^2$ 极小城市的建造过程中同样得到转换。因为城市内部的对象物尺度的设定，大多是以我感觉是否合适来决定的。如桌子的高度、椅子的高度等，是通过一个可调节高度的板子不断微调而最终确立的。适合自己的尺度，是这个 $60m^2$ 极小城建设过程中的根本坚持。

usage of sandy soil as its construction material. How d Rammed Earth Structures show or make the judgment o its dimensions? The construction process of Morocca houses reveals these "judgments" for us. In Morocco the house height, the wall depth, the door height, and th window height are all accorded to its builders. In anothe word, people decide the house dimensions based o their own body sizes. Usually people would measure th window width by their own palm width, and then decid the house height according to their own height. So tha throughout the whole construction, dimensions of th human body automatically embed into the building throug measurements.

Therefore, when you walk into these constructions or take closer look at them, you could feel the connection betwee the construction and its builder. The sense of meter an rhythm came from the human body dimension is vividl shown in the building.

Hence, the harmonic scaling relations transformed fro the human body to Rammed Earth Structures gives us th sense of beauty and coziness for the constructions. In th meantime, Rammed Earth Structures would use local sand soil as its building material. So people describe Rammec Earth Structures as the buildings growing from the ground but only with restructured scaling relations and embeddec with human body dimensions. This relation betweer humans and dimensions is also used in the $60m^2$ minima city. I tried to carefully adjust all objects' dimensions inside the city by an adjustable board, for example the height o tables and chairs. Insistence on a suitable size of one's owr

图：作为对象物而存在的“广场”中
暖气片要素

igure Left: The radiator element at
e "square"

空间与装置

在空间设计中，除上述的颜色、视线、尺度等问题之外，装置的设计也是非常重要的。空间设计中的装置，不仅仅只是一个装饰或摆设，更是一种观念的投射与表达。在这个极小城市中我们设置了如下的装置与道具。

装置 1—塔

“塔”是城市中的常见风景，但是“塔”的存在，一般情况下都是多重的。比如曾经探访过的意大利托斯卡纳地区的圣·几米尼亚诺古城，在这个中世纪小城中随处可以看到高塔，这里的“塔”是权力地位的象征。而同样意大利托斯卡纳地区锡耶纳广场上的钟塔则是那个小城当中的制高点。因此如果把家看成是一个城市的话，那么“塔”就应该具有其特殊的作用。在 60m² 极小城市中，“塔”的装置成为了具备采光功能的灯塔。这个灯塔不仅能够满足向上和向下的照明，同时其侧面打开，还可以作为存放 CD 的架子来使用，使“塔”成为一个多功能的装置。

装置 2—灯具

前面提到的“广场”侧面的“路灯”灯具，是使用很薄的三合板做成一个灯槽挂在墙壁上，并在其中嵌入灯泡完成的。这样的做法一方面使其成为一种照明器具的同时，也似乎在暗隐着中国古建筑中伸出的檩条的构造搭接关系。

装置 3—楼梯

在城市和聚落的空间中，楼梯和台阶梯段是空间中

is rooted during the whole constructing process of the 60m² minimal city.

Space and Device

In space design, along with color, sight, and dimension issues the devices design is really important as well. The devices in space design are not only furnishings and decorations, but also an expression and a reflection of a certain concept or idea. Within this minimal city, we setup the following devices and props.

Device 1—tower

"Towers" are usual scenery in cities. However, "towers" are usually characterized as multiple existences. Take San Gimignano ancient city in Tuscany region, Italy as an example.

In this medieval small town, towers are everywhere. On one hand, towers represent power and social status. The bell tower at Piazza del Campo of Siena, Tuscany, Italy is the Commanding heights of the city. When we treat home like a city, the "tower" in it should be functional in a special way. In the 60m² minimal city, the tower device becomes a lighting functional 'light house'. This lighting tower can satisfy the lighting needs for different directions and it can be functional as a CD shelf when it's opened from the side, which makes the "tower" multifunctional.

Device 2—Lamp

The "road light" lamp at one side of the "square" as mentioned before hangs on the wall with a lamp groove

右图：打开楼梯下面柜子门之后的“街道”空间的场景

Figure Right: The "street" space of th stairs with open doors

的功能性要素的同时，也是一个景观意义上的装置。不仅希腊的米克诺斯岛上聚落中存在，中国羌族聚落、藏族聚落以及河北的石头村中都存在。不同的是这些楼梯是出现在不同的空间里而已。当一个空间当中出现某种楼梯的形式,会引起一系列曾经见到的楼梯的相关的“似曾相识”的联想。而设计时将“楼梯”作为一个装置引入极小城市中的“街道”，也正是希冀这种联想的存在与扩散而将其引入的。

这个作为壁柜而存在的楼梯的引入，实际上还有另外一层想法，就是希望这个原本只是一层的住宅，让人有在其之上还存在着二层的感觉。因为一般情况下，人一看到这个楼梯,心理上立刻会有之上还有一层的假定，于是无形中使建筑的面积在观者的心理上获得了被放大了的感觉。也可以说这个楼梯是一个活跃室内视线的道具，是一件具有多样用途的装置。

装置 4—百叶窗

百叶窗是极小城市中的重要装置。首先考虑到的是百叶窗对光线的调节功能，百叶的上下翻转调节可以让室内的光线有丰富和微妙的变化。更重要的考虑是，由于所在房间的现实中的窗外，景观杂乱，令人烦躁，没有丝毫值得观赏层面的价值，利用百叶窗的叶片，可以有效阻隔外面的景象，并通过百叶对外部的景观进行横向的切割，进行片段化的摘取。此外，百叶窗更重要意义还在于，在 $60m^2$ 极小城市中营造并保持抽象的空间感觉气氛，不想因为一个透明的四方窗洞的存在而使人的视线触及外面的杂乱的具象现实世界。一旦视线落到了一个杂乱的具象物体上，一切真实的尺度就会立刻被还原，头脑中就会立刻将想象与现实进行比对，梦幻感

made of thin board and embedded light bulb in it. A meantime it serves as the lighting tool, it also implies th structural connecting relation of purlines in ancient Chines architectures.

Device 3 – Stairs

In cities and settlements space, stairs and staircase serve for both functional elements and scenery devices It's showed in Mykonos island settlements in Greece a well as in Chinese Qiang minority settlements, the Tibeta settlements and the stone village in Hebei, China. The onl difference is these stairs existed in different spaces. Whe people see one certain form of stairs in one space, th series association of stairs in other forms would come t mind. Putting the "stairs" as a device into the minimal cit carries out and spreads the idea of this association.

The stairs introduced as a closet actually has another leve of function, which is to make a one floor apartment to loo like it has a second floor.

Normally people would automatically assume there i another floor in the apartment. It gives space observers mental sense of enlargement in terms of the space. Not t mention it adds variety to indoor devices.

Device 4 – Blinds

The use of blinds is another key point in the minimal city The first consideration is lights adjustment function to th rooms. Addition to that, it blocks the annoying view outsid the windows in reality. Those horizontally expanded blind cut the outside views in a certain way. More importantly, i

左上图与左下图：“卧室”中的一角
右图：“广场”标志性建筑的鸟瞰
Figure Up Left & Down Left: both are one corner of bedroom
Figure Right: An aerial view of the landmark on the "square"

左图："广场"中的"影像与映射"
Figure Left: The "image and reflection" at the "square"

消失。

百叶窗能够使得室内空间相对隔绝。通过百叶窗对室外世界的屏蔽，使人身处一个相对封闭的小环境当中，从而内心的感受、意识中的空间感觉得以充分的释放。

装置 5—既存"暖气片"

在 60m² 极小城市的设计过程中，也遇到了不少"偶然为之"的"发现"。比如房间内原来配置的老式的暖气片，在装修的过程中，装修施工的工人师傅都曾好心地向我建议换成时髦的新式暖气片。通过"审视"，"发现"这个老旧样式的暖气片本身其实就是一件艺术品，从整个室内调性来看，它的存在，作为空间中的"对象物"非常恰当。将这个"物件"摆放在房间里，通过增加隔板，拜访几何体等运作方式，使其有一种"实用品"或是"艺术品"的暧昧倾向，产生一种模糊的氛围。

装置 6—映像

在 60m² 极小城市的"广场"上放置了一个投影设备，投影的屏幕就是那个起着分隔空间作用的"橱窗"的背面—"广场"上的"标志性建筑"。投影所带来的不同的影像内容可以变换这个"广场"空间的性格，使其变得更加富有动感。同时，通过影像的投射，让现实世界变得模糊不定，使现实与非现实之间产生错乱。当这样一个现实的空间变得抽象和非现实的时候，矛盾对立的氛围会造成别样的体验。

"广场"上的"标志性建筑"的底部还特意设计了一个突出的台子，相当于这个"标志性建筑"的大台阶，

helps the 60m² minimal city to keep and build a sense of abstract space. It prevents people from seeing the chaos in concrete reality. Once people's eyesights land on those detailed messy objects from outside, the sense of dreamy abstract space would immediately vanish in the retrace of the real world.

Blinds could make the indoor space relatively isolated. Shielding the outside world and experiencing a relatively sealing environment definitely provides people a better space sense.

Device 5 – Radiators

There are random discoveries in the 60m² minimal city design. The old fashioned radiators are one of them. Decoration workers suggested me to replace these old ones to modern new models. However, it came to me that these old fashioned radiators themselves are truly art pieces.

Their existences are appropriately served as "objects" in the space. Placing these radiators in the rooms by adding boards and geometries creates a sense of ambiguous between art objects and useful goods.

Device 6 – Images

There is a projector device placing in the "square" in the 60m² minimal city. The projector screen is the backside of the space dividing "show window" – the "landmark" in the "square". Different images from the projector reflect the changing of the "square's" personality and make the room full of motions. Also, these images blur this indoor

右图：远处的“圣几米尼亚诺的塔”
Figure Right: The "San Gimignano tower" from fa

使“标志性建筑”成为“广场”的焦点，如同城市中市民活动广场中的舞台，在视觉上制造一个趣味中心。还由于“标志性建筑”进深比较薄，台基可以让这个竖向的柜子与水平面的地面之间有一个过渡，形成一个起转折作用的基台。这个基台本身具有实用的功能，打开上面的盖子，下面是一个横向的储物柜，是一个存放图纸等需要展开存放的储存设施。

抽象与想象

几何学是人与空间进行对话的基本手段。在 $60m^2$ 极小城市的设计中，具象东西的出现被尽量地控制，所以鲜有花瓶、盆栽等这些具有丰富的装饰的东西，取而代之的是方体、球体、圆锥等几何体块。

从立体派的视点来理解，自然界所有的东西，即具象的东西实际上都可以还原成抽象的几何体块，比如三角锥、圆球、方块等。既然如此，那么反过来看的话，这些被还原的一个个纯粹的几何体块，其所对应的应该是自然界当中所有的具体和具象的事物。基于这样的思考，为了唤醒人在不同的心境、不同的季节、不同时间段中在这个空间当中行走时的丰富感受，采用简单的几何体是企图使人们能够通过其所看到的对象物获得更多、更加丰富的想象，从而规避单一的物体与我、客体与主体之间的这种一对一的信号指向。

空间经验还原器

以往虽然走了不少城市和聚落，但是并不是所有的

real world and disorder the reality and the non-reality When a realistic space becomes abstract and unrealistic a paradoxical atmosphere could give people an unusua experience.

The "square landmark" places on a platform. This makes the "square landmark" a focus. The platform is like a stage for performance and visually creates an entertainment center. Furthermore, the base of the platform serves as a transition from vertical wardrobe to the horizontal ground surface due to the fact that the depth of the "landmark" is relatively thin. It is actually useful as well. When you open it, it's a horizontal storage space and could store blueprints or something needed flat space.

Abstraction and Imagination

Geometry is a basic communication between people and space. In the $60m^2$ minimal city, concrete objects are limited in purpose. There are few decoration goods like vases or potted plants but many geometry objects such as cubes, globes, and cone.

From a cubism point of view, everything concrete in nature could be reduced back to a geometry object, such as cones, globes, or cubes etc. On contrary, all kinds of geometry objects should represent all concrete objects from our nature. Based on these thoughts and the purpose of giving people alert for different mood, seasons, and timing, the usage of simple geometries functions as a trigger for vivid imaginations when people see them. This tries to avoid the simple objects and people or subject and object signals.

斤见都深深地刻在脑子里。只有那些真正打动过我或者在脑海中留下深刻印象的那些东西才会留存，并长寸间地印在脑海里。这些存留下来的东西，是一种经佥性的积淀。这种被称为积淀的东西，是脑海中经过泽优性的选择而获得的一个结果。所谓择优性，通俗来说就是人体大脑的选择性。而这样经过择优选择出勺记忆片段，在进行设计工作的时候，就会在脑海中进行择优性的流出，就是说，当进行设计构思，勾勒脑海中理想事物时，手受到大脑控制，会在不经意间把那些曾经的美好的、印象深刻的、打动心灵的、最子的东西流淌出来。而流淌出呈现展示状态的那些场景，一定是当时看到的那些最好景象的模糊记忆。

可以说，空间的设计和空间的体验实际上是物质空间和精神空间的相互转换，在这个“互换”的过程中，“得意忘形”始终是一个关键的要点。我总是在想，人的精神是通过经验与记忆来保持持续的，物质仅仅是一时作为“唤起”记忆的载体而存在。因此我所理解的“家”更是一种可以进行自由调戏的“空间的玩具”，这种“玩具”还承载着空间的体验与空间的再次还原的功能。从这个意义上来理解，这里所展示的“60m^{2}极小城市”实质上是一个“空间经验的还原器”。

The Restoration Device of Space Experiences

Although I've been to a lot of cities and settlements, I couldn't everything I've seen. Only those touching ones would tattoo in my mind. Those I stored are selectively accumulations and optimized results. Human brain performs this optimization work. When it comes to design work, the most impressive pieces, the most soul touching pieces and the most meaningful pieces would subconsciously jump on my blueprints when I try to perform an ideal design. All the sceneries come to real have to be those best memories from the past.

At some level, space design and space experience are in fact an interchange of material world and spiritual world. During this interchanging process, it is a key to forget all formations and understand the higher level of meaning. I've always been thinking that the spiritual world of a person is formed by experiences and memories and materials exist only for "awakening" these meaningful memories. So my understanding of a home is more like an adjustable "space toy" and this "toy" stores space experiences and restoration function. Therefore, we could call this the 60m^2 minimal city a "restoration device of space experiences".

左图：将“窗”的要素闭合
Figure Left: The closing 'window' element

作者简介

王昀 博士

1985年 毕业于北京建筑工程学院建筑系
获学士学位
1995年 毕业于日本东京大学
获工学硕士学位
1999年 毕业于日本东京大学
获工学博士学位
2001年 执教于北京大学
2002年 成立方体空间工作室
2013年 创立北京建筑大学建筑设计艺术研究中心
担任主任
2015年 于清华大学建筑学院担任设计导师

建筑设计竞赛获奖经历：
1993年日本《新建筑》第20回日新工业建筑设计竞赛获二等奖
1994年日本《新建筑》第4回S×L建筑设计竞赛获一等奖

主要建筑作品：
善美办公楼门厅增建，$60m^2$极小城市，石景山财政局培训中心，庐师山庄，百子湾中学，百子湾幼儿园，杭州西溪湿地艺术村H地块会所等

参加展览：
2004年6月“‘状态’中国青年建筑师8人展”
2004年首届中国国际建筑艺术双年展
2006年第二届中国国际建筑艺术双年展
2009年比利时布鲁塞尔“‘心造’——中国当代建筑前沿展”
2010年威尼斯建筑艺术双年展，
德国卡尔斯鲁厄Chinese Regional Architectural Creation建筑展
2011年捷克布拉格中国当代建筑展，意大利罗马“向东方-中国建筑景观”展，中国深圳·香港城市建筑双城双年展
2012年第十三届威尼斯国际建筑艺术双年展中国馆等

Profile

Dr. Wang Yun

Graduated with a Bachelor's degree from the Department of Architecture at the Beijing Institute of Architectural Engineering in 1985.
Received his Master's degree in Engineering Science from Tokyo University in 1995.
Received a Ph.D. from Tokyo University in 1999.
Taught at Peking University since 2001.
Founded the Atelier Fronti (www.fronti.cn) in 2002.
Established Graduate School of Architecture Design and Art of Beijing University of Civil Engineering and Architecture in 2013, served as dean.
Served as a design Instructor at School of Architecture, Tsinghua University in 2015.

Prize:
Received the second place prize in the "New Architecture" category at Japan's 20th annual International Architectural Design Competition in 1993
Awarded the first prize in the "New Architecture" category at Japan's 4th SxL International Architectural Design Competition in 1994

Prominent works:
ShanMei Office Building Foyer, $60m^2$ Mini City, the Shijingshan Bureau of Finance Training Center, Lushi Mountain Villa, Baiziwan Middle School, Baiziwan Kindergarten, and Block H of the Hangzhou Xixi Wetland Art Village.

Exhibitions:
The 2004 Chinese National Young Architects 8 Man Exhibition, the First China International Architecture Biennale, the Second China International Architecture Biennale in 2006, the "Heart-Made: Cutting-Edge of Chinese Contemporary Architecture" exhibit in Brussels in 2009, the 2010 Architectural Venice Biennale, the Karlsruhe Chinese Regional Architectural Creation exhibition in Germany, the Chinese Contemporary Architecture Exhibition in Prague in 2011, the "Towards the East: Chinese Landscape Architecture" exhibition in Rome, the Hong Kong-Shenzhen Twin Cities Urban Planning Biennale, Pavilion of China The 13th international Architecture Exhibition la Biennale di Venezia in 2012.

图片来源；
以下页码中的图片均为周之毅摄影
文前06页
以下页码中的图片均为方振宁摄影
页码：文前08页、02、20、23、24（上）、29、30、32、34、40-42、45（上）、46、48、49、51-59、61、62、65-68、71、72

其他图片如无特殊说明均为作者本人提供

图书在版编目（CIP）数据

60平米极小城市 / 王昀. -- 北京 ：中国建筑工业出版社，2015.5
ISBN 978-7-112-17977-0

Ⅰ. ①6… Ⅱ. ①王… Ⅲ. ①城市—住宅—建筑设计—研究 Ⅳ. ①TU241

中国版本图书馆CIP数据核字（2015）第062022号

责任编辑：徐　冉
责任校对：李欣慰　关健
封面设计：赵冠男
版式设计：赵冠男

感谢北京建筑大学建筑设计艺术研究中心建设项目的支持

60平米极小城市
王昀
*
中国建筑工业出版社出版、发行（北京西郊百万庄）
各地新华书店、建筑书店经销
北京顺诚彩色印刷有限公司印刷
*
开本：787×960毫米 1/16 印张：5¼ 字数：129千字
2015年6月第一版　2015年6月第一次印刷
定价：52.00元
ISBN 978-7-112-17977-0
(27198)

本书的研究及编辑出版工作得到了以下人员的大力支持:
李晓红、高媛、赵子宽、孙炼、张莹莹、王子源、楼洪忆、
吴伟、刘艳羽、刘月莉、李德龙、刘海舰、王遐、
朱丹青、李孝慈、张君忆